建筑立场系列丛书 | No.55

灰色建筑中的绿色自然：
混合型建筑设计

Green in Grey
Architecture in Hybrid Mode

汉英对照
（韩语版第371期）

韩国C3出版公社 | 编

于风军 安雪花 汪冉 焦明 杜丹 孙探春 | 译

大连理工大学出版社

灰色建筑中的绿色自然
混合型建筑设计

- 004 *灰色建筑中的绿色自然：混合型建筑设计* _ Angelos Psilopoulos
- 010 科学与生物多样性小学 _ Chartier Dalix Architects
- 022 巴黎东区科学与技术区 _ Jean-Philippe Pargade Architecte
- 038 植被25建筑 _ Luciano Pia
- 054 绿化改造 _ Vo Trong Nghia Architects
- 060 猎鹰总部二期 _ Rojkind Arquitectos + Gabriela Etchegaray
- 070 Point 92建筑 _ ZLG Design
- 082 垂直森林 _ Boeri Studio

城市住宅
流行化和大众化

- 092 *流行化和大众化* _ Alejandro Hernández Galvez
- 096 阿尔比大剧院 _ Dominique Perrault Architecture
- 108 什切青新爱乐音乐厅 _ Barozzi/Veiga
- 118 城市文化欧洲区域中心 _ Atelier d'architecture King Kong
- 128 圣马洛文化中心 _ AS. Architecture Studio
- 134 瓦莱塔城门 _ Renzo Piano Building Workshop
- 150 塔德乌什·坎特CRICOTEKA博物馆 _ Wizja sp. z o.o. + nsMoonStudio
- 166 格旦斯克莎士比亚剧场 _ Rizzi-Pro.Tec.O

- 180 建筑师索引

Green in Grey
Architecture in Hybrid Mode

004 *Green in Grey: architecture in hybrid mode* _ Angelos Psilopoulos

010 Primary School for Sciences and Biodiversity _ Chartier Dalix Architects

022 East Paris Scientific and Technical Pole _ Jean-Philippe Pargade Architecte

038 25 Green _ Luciano Pia

054 Green Renovation _ Vo Trong Nghia Architects

060 Falcón Headquarters 2 _ Rojkind Arquitectos + Gabriela Etchegaray

070 Point 92 _ ZLG Design

082 Vertical Forest _ Boeri Studio

Urban How
Popular and Public

092 *Popular and Public* _ Alejandro Hernández Galvez

096 Albi Grand Theater _ Dominique Perrault Architecture

108 Philharmonic Hall in Szczecin _ Barozzi / Veiga

118 Euroregional Center of Urban Cultures _ Atelier d'architecture King Kong

128 Saint-Malo Cultural Hub _ AS. Architecture Studio

134 Valletta City Gate _ Renzo Piano Building Workshop

150 CRICOTEKA Museum of Tadeusz Kantor _ Wizja sp. z o.o. + nsMoonStudio

166 Gdansk Shakespearean Theater _ Rizzi-Pro.Tec.O

180 Index

灰色建筑中的绿色自然

Green in Grey
Architecture in Hybrid Mode
混合型建筑设计

只要人类在寻求建筑物的庇护，大自然就会是建筑设计的主要背景。无论是原始茅屋中蕴含的风水（宇宙）论，还是古典装饰遵循的自然规律，都显示出"自然"便是本真，亦或是完美。尽管人类致力于从不可脱离干系的"自然"家园中逃脱出来，但是大自然仍保持着它与人类的联系，尽管有些与人类智慧提出的要求相悖。在工业时代，技术被定义为违背自然的人造产物，这个过程和结果便造就了现在的世界，一个人类用短期的获利——经济利益，来证明自己存在的世界。这就像是现实生活中的巴氏消毒法[1]，即人类与细菌的战争，视自然为潜在的不健康因素。于是，人工技巧成为了一种新的设计准则，自然便被改造成公园、动物园、建筑群和其他类似的人类成就。

直到21世纪初，这一切发生了显著的变化。碳排放导致的温室效应、森林乱砍滥伐、物种灭绝等问题，被认为与反人类的直接犯罪行为一样严重。在现今发达社会，这些环境问题对人类的影响是毋庸置疑的，而对于这些环境问题的呼吁，如同经济利益问题，也不容忽视。

在当今时代，科技并非授予人类驾驭自然的权力，而是帮助重建人与自然的共生条件，而这需要更高层次上的战略。建筑物本身爱莫能助，建筑只能以人类期许的新的形式呈现。现在的建筑，既没有失去可以让人身心愉快的自然元素，又拥有强大的混合式机器的能力，利用自然科学和人工技术的融合，创造出可以自给自足的居所，重新带来微观和宏观的气候变化，提供富有矿物质的天然氧吧环境。最重要的是，这样的建筑可以成为未来建筑设计的原型，在设计中引入更多的变化元素，比如最明了却也最具有野心的设计理念，即重建城市生态环境，超越城乡区划，模糊城乡边界，融合人工与自然。

Nature has served as a prime reference as long as man has tried to shelter himself in building. From the cosmology of the primitive hut to the natural order of classical ornament, to be "natural" would practically mean to be true, or ideal. As humanity tried to emancipate itself from the inevitability of its "natural" habitat, nature still remained relevant even in the form of something opposite to which the human intellect would make its claim. In the industrial era technology is defined the artificial in defiance of the natural. The quantification of processes and outcomes created a universe where man established his presence with short-term gain – e.g. financial profit. Similarly the "pasteurization" of everyday life[1], namely humanity's battles against germs, saw nature as something potentially unhealthy. Artifice became the new norm, and nature was to be tamed in the form of parks, zoos, collections, and other similar human achievements.

By the dawn of the twenty-first century things have resolutely changed. For one thing, issues like global warming through carbon emissions, deforestation, the extinction of species, etc. are considered as critical as direct crimes against humanity, their impact hardly being disputed and their appeal as topics of interest hardly being neglected by advanced societies.

In our day it seems as if technology is used not to empower man over nature but to re-establish the terms of their symbiosis. While this involves strategies on a hugely larger level, architecture cannot help but position itself in the new norm. Without forfeiting the pleasurable aspect of nature, buildings nowadays stand as powerful hybrid machines, utilizing nature and its ways in fusion with man-made technology to create self-sustaining habitats, micro- and macro- climate regenerators, and powerful containers of minimal energy footprint. Most importantly they stand as prototypes, one building serving as a model for a wider adaptation, with the most apparent – as well as ambitious – scope of negotiating the terms of a new urban ecology, transcending divisions and blurring the lines between urban and rural, man-made and natural. Ultimately this becomes a problem of awareness, and the buildings in our survey are exemplar at exactly that aspect.

科学与生物多样性小学_Primary School for Sciences and Biodiversity/Chartier Dalix Architects
巴黎东区科学与技术区_East Paris Scientific and Technical Pole/Jean-Philippe Pargade Architecte
植被25建筑_25 Green/Luciano Pia
绿化改造_Green Renovation/Vo Trong Nghia Architects
猎鹰总部二期_Falcón Headquarters 2/Rojkind Arquitectos + Gabriela Etchegaray
Point 92建筑_Point 92/ZLG Design
垂直森林_Vertical Forest/Boeri Studio

灰色建筑中的绿色自然：混合型建筑设计
Green in Grey: Architecture in Hybrid Mode/Angelos Psilopoulos

对棍，1820–1823，Francisco Goya
Fight with Cudgels, 1820~1823, Francisco Goya

　　建筑师一直着迷于大自然的魅力。Vitruvius在其著作《建筑十书》[2]中全面阐述了大自然在建筑中的重要角色，比如运用自然的风向、气候、土壤和水源，都可以在建筑中起到重要的作用。同样，大自然也提供了理想背景，就像是个鬼斧神工的建筑师，将人类的居所变成了大自然中的基本的元素。尽管如此，人类一直在孜孜不倦地提高其基本要求，从功能实用性到艺术美学。而在这方面，人类似乎是在逆自然而行，人类不再视大自然为必要条件，而将自然理解为一种可以被改造成人类可适的环境，而人类的这一切努力，最终只不过是幻想而已。

　　建筑学与大自然的互动形式，就如同在历史进程中人类话语赋予建筑的意义一样多元化。[3] Sörlin[4]界定了自然在建筑学中的作用，或是作为建筑中的构成要素，亦或是建筑中的装饰主题：从流动空间和曲线表面，到动物和植物，螺旋结构和藤蔓花式，到系统的建筑理念，即功能性建筑（如公园或动物园），或是原型的背景（如经典几何图案的"自然"法则，或是天圆地方宇宙论决定的圆形的帐篷、茅屋或蒙古包）。即便如此，建筑学还是不能视自然于不顾，单纯追求"纯粹概念空间"，而全然依赖人类智慧，背离建筑与自然共存的理念。从广场设计或轴向布局，到基于程序生成的设计原理，建筑学探究的不仅是单纯模仿自然，而是分析并利用自然规律，从而建立人类理想的与自然共生的空间。

　　这些就如同是一枚硬币，自然是硬币的核心，而人类的活动则是刻在硬币表面上雕花。因为自然科学博大精深，因此大自然对人类如何运作不屑一顾。自然尽力维持着其内部的精确平衡，无论人类如何表现自我，来维持其地位。Michel Serres的《自然契约》[5]的开篇，描绘了Goya的一幅画，画中是"两个陷在流沙中的人，他们手持棍棒在争斗"。在构图上，Goya将这两个人放在画的正中央，"这两个人的膝盖已经深陷在淤泥中，而且他们每动一下，就会陷得更深一些，最终，他们将葬身于淤泥之中。他们下陷的速度取决于他们斗争的激烈程度：斗争越激烈，动作越

Architect has always been fascinated with nature. Vitruvius has written extensively in *De Architectura*[2] about the fundamental role nature plays in architecture, mastery of winds, climate, soils and water being the very condition for the act of building. In quite similar terms, nature has provided for archetypical references, being itself "a builder" of sorts, in as much as the act of shelter becoming a most fundamental part of architecture. Nevertheless man has strived to elevate his base needs, covering a range from utility to the very extreme of aesthetic appreciation. In this position man can be thought of counteracting to the inevitability of nature; he no longer thinks of it as a necessary condition but rather as a context that can be formulated to suit his survival needs, then his endeavors, become ultimately his fantasies.

In these terms architecture has responded to nature in ways as numerous as the meanings that were assigned to it in the history of human discourse[3]. Sörlin[4] basically distinguishes nature in architecture for its role either as a formative element or a decorative theme: from flowing spaces and curving surfaces, to animals and satyrs, helixes and vines, to the very programmatic principle of architecture, namely a function(e.g. a garden or a zoo) or an archetypical reference (e.g. the "natural" principles of classical geometry or the cosmological reflection in the round shape of a camp, hut, or yurt). In this context one can also not escape the mirror claim for a "purely conceptual space", one that negates the necessity for reference to nature, a sort of counter-juxtaposition to the founding principle by means of human genius. Covering a range from square form or axial arrangements to the empowerment of generative design principles based on programming, architecture thus sought not to imitate nature but to analyze its rules and supersede it, looking to create coherent universes that appeal to the ideal of human achievement.

Still, these are aspects of the same coin; only, nature should be thought of as the substance of this coin, while human activity is merely engraving its sides. As the natural sciences know too well, nature is largely indifferent to human activity. The former will remain consistent to sustain its elaborate equilibrium regardless of the various narratives that man assumes trying to maintain his position in the picture. In the opening chapter of his book "The Natural Contract"[5] Michel Serres evokes a painting by Goya where "[a] pair of enemies brandish sticks [are] fighting in the midst of a patch of quicksand". While the composition of the painting puts the duel of the two men center stage, Goya "has plunged the du-

巴黎东区科学与技术区，马恩拉瓦莱，法国
East Paris Scientific and Technical Pole in Marne-la-Vallée, France

野蛮，他们下陷的速度就越快。交战中的两个人并没有意识到他们即将陷入深渊，而作为旁观者的我们，却看得清清楚楚。"6

Serres接下来便提出关于这幅画的"一般性问题"：两个人中谁会死？谁会赢？这时最好问作为旁观者的自己，我们站在什么角度去看这幅画里挣扎的两个人的，更重要的是，我们如何看待这个问题。现在，全球变暖、工业污染、两极冰川融化等环境问题都已成为公众关注焦点，1993年讨论会中迫在眉睫的环境问题已然成为严酷的现实。在建筑学领域中，全新评价建筑样式、城市规划和土地管理的方法已然出现。现在，重中之重是何谓"可持续发展"7。这个概念主张保持平衡的概念，弱化工业社会中过度商品制造（这也可能是不得已而为之？）。可能这会是21世纪的一个新起点，在结果被迅速资本化之前，首先建立起系统的思想纲领。尽管建筑仍然代表了"具有可持续发展原则的空间"8，但新型建筑却致力于创建一个交互界面，将人类的期许和人类以更广阔的全球化视角解读的自然环境完美结合。

我们调查中的案例都是"混合型"的建筑，这些建筑超出了利用"自然"作为形式参考或装饰的做法，建筑师将自然元素与人造材料相结合，从而营造一种自然氛围和自然景观，或设计出能够调节气候的装置。同时，案例中的建筑物还具有更广泛的影响，它们将引领未来建筑学的发展潮流，其影响也可以从经典的建筑案例中感知，即让观众能够学习并坚信自己承担的社会责任，要在全球背景下来考虑。混合型建筑超越了城市与乡村、景观与建筑、强势与弱势的界限——就如同打破了Serres画中两个决斗者的对立关系，从而打造了一种微观生态系统。被赞誉为"引以为荣""梦幻之地""醍醐灌顶"的新型建筑，这些建筑重塑城市环境，重新规划了破败的环境所带来的紧张关系。

Chartier Dalix建筑师事务所设计的科学与生物多样性小学位于城市区域的狭窄且独立的地块中，这座学校将自然景观引入城市，同时自然景

elists knee-deep in the mud. With every move they make, a slimy hole swallows them up, so that they are gradually burying themselves together. How quickly depends on how aggressive they are: the more heated the struggle, the more violent their movements become and faster they sink in. The belligerents don't notice the abyss they're rushing into; from the outside, however, we see it clearly."6

Further on the chapter Serres suggests that the "usual question" is who will die, and who will win. Better ask yourselves, where are we standing as we're reading this passage and –more importantly– how. By the time subjects like global warming, industrial pollution, and the melting of polar ice became central to public awareness, and this argument of 1993 becomes a matter of pressing reality. In the realm of architecture, a new way of looking at buildings, city planning and land management in general have emerged. The principal focus now is on what was termed "a sustainable course of development"7, a concept that is founded on establishing equilibrium rather than on the ever-growing commodification of processes and artifacts our industrial society (unwillingly?) put first place. Perhaps this marks a new point of departure for the twenty-first century, establishing programmatic thinking before an outcome that is immediately capitalized. In spite of an architecture that still stands "remarkably void of sustainable principles"8, new buildings aim to act as an interface between human aspiration and "natural" adaptation to a wider, more global understanding of their environment.

The cases we examine in our survey may be seen as "hybrids". Extending beyond the utilization of "nature" as formal reference or ornament, they aim to fuse it with man-made materials in order to design atmospheres, landscapes, or machines that manipulate the climate of a place. Furthermore they submit entities that bring impact on a wider scale, taking part in an architectural trend of greater significance, its impact being perceivable in paradigm; namely in terms of educating an unsuspecting audience on the benefits – and also the responsibility – of thinking in a more global context. This hybrid architecture transcends dichotomies between city and country, landscape and building, major and minor – Serres' duelists. It produces micro-ecologies. Descriptions such as "places to be proud of", "places to dream", "places that teach", places act as urban regenerators or repurpose the tensions of a failing environment.

The Primary School for Sciences and Biodiversity by Chartier Dalix

猎鹰总部二期，墨西哥城，墨西哥
Falcón Headquarters 2 in Mexico City, Mexico

垂直森林，米兰，意大利
Vertical Forest in Milan, Italy

观对城市起到保护的作用。学校由一层石材立面覆盖，建筑内包含多层环境系统，将绿色植物与钢筋水泥完美融合。学校的目的是寓教于乐，将诗情画意与教书育人和自然再次结合。作为学校，它将生态多样性带回到城市环境的中心。从整体上看，人们最终使建筑充满了生机与活力。在校的孩子们会在其中发挥他们的潜能，而当地的居民，因这座建筑的功能和强烈的存在感，把这所学校当作"社会枢纽"。总而言之，这所学校就像是一个自给自足的生态系统，具有强大的教育功能，植根于城市的中心。

Jean-Philippe Pargade建筑设计事务所设计的巴黎东区科学与技术区，是另一个将人工空间的塑造融入自然景观的外形的建筑典范。采用法国传统的桥梁施工技术9，建筑将大跨度的绿色屋顶之下的空间最大化，并将建成的表皮融入到人工建成的"自然"水平线内。这个项目体现了环保意识，致力于被动节能技术，实现最节能的方案。绿色屋顶以一种最根本的语言形式，将场地内各分散的结构统一在一起，既实现了丰富的视觉效果，又提供了功能性休闲空间。

Rojkind建筑事务所和Gabriela Etchegaray事务所设计的猎鹰总部二期，位于墨西哥城的绿色庭院，是为了补充猎鹰总部一期办公区而修建，与一期建筑反向而建。这两座建筑都把自然元素有机融入到一个有独特设计理念的混合式形式中。猎鹰一期的黄色玻璃盒子结构的表面映射着自然，自然也成为室内能够观赏得到的散碎风景。猎鹰二期则略有不同，从理念上来讲就是一期花园的一个延续，用一个透明玻璃幕墙罩住形成一处内部空间，幕墙采用510多个组合式的盆栽进行线性装饰。在建筑内部，这些盆栽起到了遮阳伞的作用，而在建筑外部，这些盆栽主要作为建筑周围茂密的花丛的视觉延续。这两座建筑都将自然元素融入其中，作为中介，在无形中对整体体验起到重要的作用。猎鹰二期也是一个充满生机的有机体，时间会改变它的容貌，相比猎鹰一期黄色玻璃盒子的耀

Architects is set in an urban enclave that "stands apart", introducing a natural landscape into the city at the same time it is protecting it. Wrapped by a stone facade the building encompasses a multi-layered environment in which greenery and concrete are blended in shared terms. As a school, it aims to reconnect poetry and education with nature; as a feat of architecture, it aims to reintroduce biodiversity to the heart of its urban context. As a whole, people ultimately animate it. Local children will "go to fulfill their potential" while local residents will use it as "a social hub", as the building's very function and presence will ensure it. All in all it looks to act as a fully blown ecosystem, planted in the heart of the city, educative in purpose.

East Paris Scientific and Technical Pole by Jean-Philippe Pargade Architecte is another building that draws from the idea of shaping artificial space into the form of a natural landscape. Using techniques from the construction of bridges – a very French legacy[9]– it maximizes space under its long-spanning green roof and dissolves the built envelope into an artificially "natural" horizon. The project is environmentally conscious, engaging passive energy saving technologies and techniques to establish a minimum energy footprint. In a gesture of a very fundamental architectural language, the green roof unifies the disperse structures on the site, offering both a rich visual stimulus and a functional recreation area.

Falcón Headquarters 2 by Rojkind Arquitectos and Gabriela Etchegaray is a complement and at the same time a reverse mirror to the company's first installment of office space on a green courtyard in Mexico City. Both of the buildings fuse the natural into a hybrid form of distinctive conceptual merit. In Falcón 1, nature can be seen as a reflection on the yellow glass surfaces of its box-shaped outer shell as well as a series of fragmented views from the inside. In subtle contrast, the Falcón 2 was "conceptualized as an extension of the garden itself", defining its interior volume with a transparent glass curtain wall that is distinctively marked by no less than 510 modular planters in linear shape. These planters serve as sunshades from the inside, yet they function prominently from the outside as a visual extension of the dense flora that surrounds the building. Both of the two buildings' exercises integrate the natural element as an intermediate interface, almost in intangible yet fundamental to the overall experience. Still, Falcón 2 does not fail to also act as a living organism, growing and changing its face in the passing of time, establishing presence by hiding rather than standing out as its yellow glass box counterpart does.

植被25建筑,托里诺,意大利
25 Green in Torino, Italy

照片提供:©Beppe Giardino

眼,它显得更加隐蔽和低调。

Luciano Pia设计的植被25建筑,毫无疑问是建筑界的一个异类。这种建筑利用叙述性,而非外形来唤起一种奇妙的体验。其功能理念非常简单:Pia想通过在住宅单元和街道之间建造一个过滤器,来形成"一个内外之间的平坦的过渡区"。通过设计,这个过滤器的目的是实现室内外的新陈代谢;钢梁、桁架和支柱组成的三维框架成为一系列嵌入构件(包括曲形阳台以及最重要的植物)的支撑框架。人工与自然的融合,不仅仅是位置上并置一起,还要实现形式上的融合。支柱变成金属桁架,将公寓转变为融于森林的三座房屋和阳台,且带有超大号的花盆,而这些花盆用来种植一些树木,而不仅仅是装饰性绿植。进入到这座复杂的结构犹如进入一座乐园,使人们回忆起儿时时光。同时,建筑也在默默地发挥控制微气候和宏观气候的作用,节约能源,在原有的城市工业区中建立生物多样性。尽管Pia并非一直使用夸张的建筑语言,但植被25建筑确实独立于周围环境中,以将我们的生活融入了某些梦幻的元素。

越南Vo Trong Nghia建筑师事务所在河内设计的绿化改造项目,是在其重新战略性地调整建筑周围破败的环境规划后脱颖而出的。例如,防盗栏换成了镀锌的钢骨网格,能够让绿色藤蔓植物生长,或者是架高地面来防止潮气和湿气。建筑的立面装饰成"绿色瀑布",这样就形成了双重视野:从建筑里面欣赏到的是森林一样的景象(在典型的城市社区),从外面看,这座布满绿叶的墙壁在这片单调的区域非常吸引人的眼球。同时,这样的立面可以遮挡住西面的阳光,而从重新设计的楼梯间的天窗里射入的自然光洒满室内,天窗同时也是一个自然的通风口。显然,这座建筑是一个设计原型,其方案可以应用到很多热带城市的典型房屋中。这个项目使用一个朴实的方式,再次引入与自然共存于城市中心的所带来的益处,并以自己为例来证明这种再生方式是简单可行的。

ZLG设计的Point 92建筑将垂直分区的复杂设计理念付诸实践,从

25 Green by Luciano Pia is undoubtedly a strange beast. Utilizing the means of narrative more than mere form and geometry, it evokes an experience that is fundamentally uncanny. The programmatic principle is quite simple: Pia means to facilitate "a smooth transition between the inside and the outside" by installing a filter between the residential units and the street. By design, this filter is a metabolist's dream; a three-dimensional grid of steel beams, trusses and columns, stands as the supporting frame for a number of plug-in elements including curve-shaped balconies and – perhaps most imposingly – plants. The artificial blends with the natural, not only in juxtaposition but also in form. The columns turn into metal trusses, transforming the apartments into tree houses and the balconies into forests, with the oversized planters used to plant trees rather than typical decorative greenery. Entering the complex structure is an act of entering into wonderland, evoking a child-like response to an environment. At the same time, the building will work silently in controlling micro- and macroclimate, saving energy and establishing biodiversity in a former industrial part of the city. In spite of the fact that the extravagance is not always Pia's architectural language, 25 Verde stands absurdly distinctive within its surroundings in order to infuse our lives with something of a dream.

Green Renovation residential project in Hanoi by Vo Trong Nghia Architects stands out first and foremost as a strategic repurposing of the very building's failing conditions; for example, security bars were exchanged with a galvanized steel trellis where green climbers would grow, or the ground floor was elevated to prevent rising damp and condensation. The facade, dubbed as a "greenfall", allows for a two-fold view: one from the inside of the residence, where the owners enjoy a forest-like image (in the place of a typical urban settlement), and one from the outside where the leafy wall acts as an attractor to an otherwise dull area. While this facade acts as a shade against punishing west sunlight, the more filtered natural light bathes the interior through a thoughtfully placed skylight at the redesigned staircase, also working as a natural ventilation duct. Evidently the building is considered as a prototypical design, its scheme being applicable to many typical houses in tropical cities. In reintroducing the benefits of living with nature in the heart of the city, the project is, at the same time, a modest approach. As such, it is indicative of a stance that exemplifies the simplicity of means needed for this kind of regeneration.

Point 92 by ZLG Design puts in operation an intricate idea of verti-

1. Cf. Bruno Latour, *The Pasteurization of France*, Cambridge, Massachusetts: Harvard Univ. Press, 1993.
2. Vitruvius Pollio, *The Ten Books on Architecture*, (1st century BC).
3. Sörlin ("Nature", in Arne Hessenbruch[ed.], *Reader's Guide to the History of Science*, [Fitzroy Dearborn: London; Chicago, 2000], pp. 122~127) recounts Lovejoy's("Nature" as Aesthetic Norm, Modern Language Notes, Vol. 42, no. 7, November 1927, pp. 444~450) claim that "nature" "had approximately sixty established meanings [and that] this number has not diminished since"("Nature", p. 123).
4. Sörlin, *Nature*, p. 124.
5. Michel Serres, *The natural Contract*, Ann Arbor: University of Michigan Press, 1995.
6. M. Serres, *The Natural Contract*, p.1.
7. See World Commission on Environment and Development(ed.), *Our common future*, Oxford, New York: Oxford University Press, 1987), p. 16; also known as *The Brundtland Report*.
8. Sörlin, *Nature*, p. 126.
9. [9]See: École des Ponts ParisTech [École Nationale des Ponts et Chaussées] established in 1747, a prestigious "Grande École" founded on the study of the building of bridges, canals and roads, and the training of civil engineers. www.enpc.fr
10. Boeri Studio, Vertical Forest, http://www.stefanoboeriarchitetti.net/en/portfolios/bosco-verticale/ Accessed 28/5/2015.

地理位置选择，到塔楼类型方案，采用了一个意想不到的、独具匠心的方案，从而设计了独一无二的建筑。建筑师在整体构架上将这座建筑分为三部分：第一部分是植被覆盖的倾斜的停车场底层，第二部分是位于建筑中层的两层透明大堂，第三部分是高层的白色混凝土办公大厦。纵观这个设计的各种形式和技术，自然与建筑的有机融合可以体现在三个层次：其一是底层的隐蔽覆层区，第二是中间的市民购物中心，第三是上层露台（看起来是从主体量中切割出来的）的主展览区。和本篇文章中提及的其他建筑一样，显然，Point 92建筑如同一台绿色机器在运作，这个设计的亮点在于在所在地理位置和周围环境内如何使这样的宏伟建筑设计真正实现与自然和谐共处。

最后，Boeri工作室建筑设计的垂直森林项目似乎总结了我们在这篇文章中讨论过的每一座建筑的特点。它位于米兰古老的新门地区，由两个塔楼组成，有约400座私人公寓，绿色植被和露台错落设计，与公寓相互连通。就像建筑师们所说[10]，这两栋楼对于大型雄心勃勃地"在城市建造森林"的项目来说，也是个原型范例。超过2000种不同的植物种植在露台上，有小型的灌木也有大型树木，相当于7000m²的森林。这些自然元素不仅能直接改善居住环境，还可以使更大范围的城市变得更宜居。垂直森林项目使用现有的科学技术解决了建筑中高层建筑的结构负重和操作问题，巧妙借鉴了高层建筑的核心理念（即面积最小化和密度最大化），利用城市与自然失去联系所使用的理念，来建立了一个新的城市自然生态景观。

cal zoning, responding in a rather unexpected and creative manner to the particularities of both the geography of the site and the typology of a tower building. In this framework the architects propose a building that develops in three parts: a green clad parking space base that blends with the sloping site, a two story transparent arrival hall in the middle, and a taller volume dedicated to office space envisioned as a white concrete block. Looking beyond the building's various formal and technological statements, nature is integrated in a similar three-fold manner: an almost camouflage-like cladding in the base level, a civic plaza in the middle, and a prominent exhibit area in the terraces that are seemingly carved out of the main volume on the upper level. As is the case with all the other buildings in our survey, the building obviously functions as a green machine. Yet, it seems that the principal merit of the design is how a building of such magnitude can show concern for co-existing with the natural tensions coming from the site and its surroundings.

Finally, the Vertical Forest project by Boeri Studio seems to sum up every single argument we made in this introduction. It consists of two towers inserted into the historic Porta Nuova district of Milan, containing no less than 400 condominium units, which are articulated in a series of interlocking green terraces and apartments. By the word of the architects[10], the two towers also stand as prototypes for a largely ambitious project of "urban reforestation". Over 2000 plants are used for the terraces, varying from small shrubs to large trees, the equivalent of a 7000m² area of forest. At this level, the natural element not only contributes to the immediate environment of the residences, but also practically stands as a terraforming machine for the larger part of the city. Using the available technology to cope with the high structural loads and operational needs, the towers themselves take the core idea of the high-rise (minimum footprint and maximum density) to establish a new urban ecology using the same principles by which our cities became almost devoid of any relationship to nature. *Angelos Psilopoulos*

Drawing©Boeri Studio

这个项目展现了利用一种混合设计理念来建造的学校和体育馆，同时也包含了第三个设计元素：促进生态多样性。从环保角度来讲，这座小学的设计采用了一种非常创新的建筑设计方案，设计理念根据其原有景观的发展，因地制宜地来绘制其纹理和设计构件。这项工程也标志着一个全新的建筑设计趋势，即生态多样性回归到城市中心地区。与其他类型的设计项目相比，学校设计是一个能激发设计师的美学灵感的项目，设计学校要重新考虑怎样的设计理念可以将诗情画意、教书育人和大自然融为一体。而这座学校的建筑设计迎接了这个挑战，将学校重建成一座功能完善的自然生态园，当地孩子能够在这里充分学习，开发潜能，且当地居民也将其视为社会枢纽。

这个设计主要包括两个部分：带18间教室的学校（其中7间学前班教室，11间小学教室）和一个体育场，这个体育场对当地居民开放。这所小学位于布洛涅-比扬古古老的场地，如今的布洛涅-比扬古已然建筑成群。设计的两个结构覆盖一层"矿物墙"，将学校和公共体育场组合在一起。

建筑主体也包括两部分：矿物覆盖区域——外立面，以及植被覆盖的屋顶。建筑的外围护结构将学校包裹起来，形成一个轮廓平滑，线条柔和的大型体量，同时体现出建筑内部的流线设计空间和建筑外部的可延展的空间，使体量间很好地结合起来。高度紧凑的建筑主体朝向当地社区，展示其高大夯实的一面。操场是两个可以相互连通的室外空间，彼此可以看得清清楚楚（总有一天学前班的孩子会升级进入小学）。建筑整体庇护了一处原生态的自然环境，而这处环境也将促进在更大型场地中心建设长期自然生态化的进程。可以说，这座建筑是有生命的，因为它会成长变化。虽然它只是一个微型的自然景观，但是在未来的五年或十年里，它将以新的外围护结构形象呈现在世人面前，就如同大自然的变幻莫测。

这个设计的亮点是"生动的"墙体，主体成分是预制的混凝土体块，这些体块展现了两种不同的纹理。可见的一面是经过抛光处理过的，平滑，可以反射光线，另一面是棱纹的，纹理粗糙。这两种纹理的墙面有助于水流快速流向建筑一侧，避免水流向墙体可见的一侧，从而有效避免墙体提前老化。这两种纹理的对比能够突出立面的厚重感，强化其浮雕感。在墙体稍矮一点的地方，高度大概2m左右，独立的墙体被设计成平滑的或者是向外倾斜的，以抵挡外来人员进入，或者可能的动物入侵。墙壁一侧的凹槽有助于植被的生长（凹槽适合蕨类植物，粗糙的混凝土适合苔藓）；而墙上的小型孔洞和褶皱结构是给动物预备的（悬垂的结构可以让燕子筑巢，角落里的空隙适合昆虫生活），同时也可以吸引多种鸟类来筑巢。

科学与生物多样性小学
Chartier Dalix Architects

©P. Guignard SAEM Val de Seine Aménagement (courtesy of the architect)

西南立面
south-west elevation

东南立面
south-east elevation

东北立面
noth-east elevation

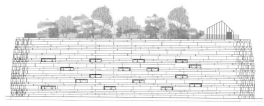

西北立面
north-west elevation

项目名称：Primary School For Sciences and Biodiversity
地点：Boulogne-Billancourt, France
建筑师：Chartier Dalix Architectes
合作者：structure_EVP, fluids_CFERM, economist_F.Bougon, HEQ_F.Boutt, ecologist_AEU, biodiversity_Biodiversita
承包商：SAEM Val de Seine
用地面积：5,164m² / 有效楼层面积：6,766m²
设计时间：2011—2012 / 施工时间：2013—2014
摄影师：
©David Foessel (courtesy of the architect) - p.10, p.12, p.18, p.20
©Cyrille Weiner (courtesy of the architect) - p.11, p.16~17, p.21

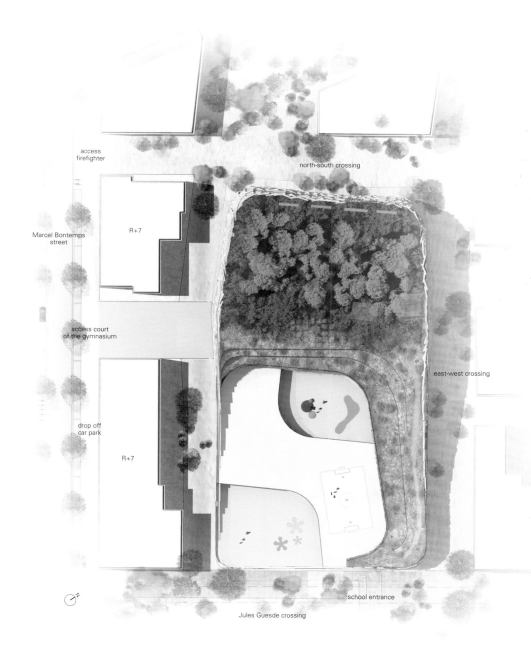

屋顶则是一座真正的空中花园，高出体育馆12m，同时也是三层不同植被的共同家园：生长在50cm深的土壤里的草地、草地外围的灌木丛，以及种在1m深土地里的树林。在这里，植物使生态走廊富有连贯性，并且各个物种在此能够相生相息。空中花园有两个功能：促进生物墙内的物种多样化和生物自身的多样化，因为花园内有丰富的资源（额外的栖居地、营养介质等），因此生态多样化得以保证。

Primary School for Sciences and Biodiversity

The project presents a mixed program to build a school and a gymnasium, but also incorporate a third element: encouraging biodiversity. It has been designed as a particularly innovative program, environmentally speaking. The concept of the building relies on the development of a primary landscape which would draw its textures and components from the landscape in which it is set. This project may well signal the start of a new trend: striving to return biodiversity to the heart of urban areas. More than any other project, building a school is an opportunity to rethink the fundamental conceptual connections between poetry, education and nature, drawing inspiration from new aesthetic impulses. Thus, the building takes up the challenge of recreating a fully functional ecosystem as a place of learning, a space where local children will go to fulfill their potential, but also a social hub for local residents. The project involves two structures: a school with eighteen classrooms (seven preschool, eleven primary school) and a gymnasium which will be open to local residents. It is located in the old Boulogne-Billancourt, now a densely built area. The two structures are united in a single volume, bounded by a same skin: the mineral wall.

There are two distinct parts to the building: a mineral section – the facades – and a section made of plants, the roof. This envelope wraps itself around the school, a general volume with smooth contours and supple lines, revealing fluid interior spaces and elastic exterior ones, avoiding ruptures between volumes. The highly compact building opens onto the neighborhood, offering a multitude of perspectives. The playgrounds are two outdoor spaces in conversation, in plain

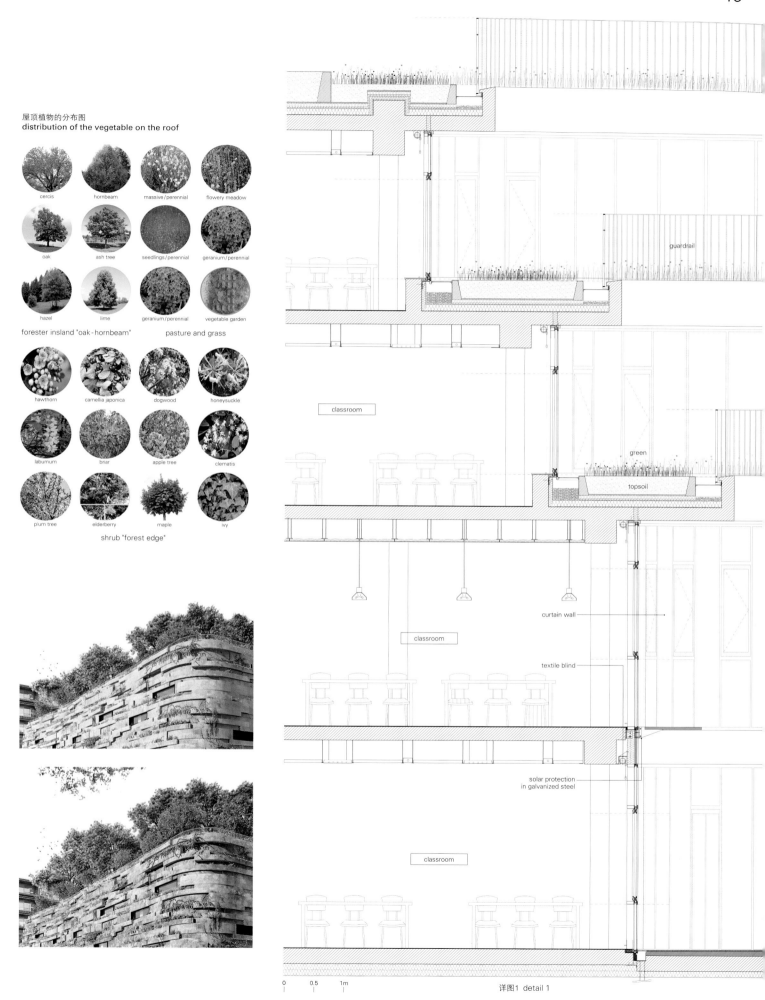

体育综合设施 ■	sport complex
学前班 ■	nursery school
小学 ■	elementary school
学前班的娱乐中心 ■	nursery school's recreation center
小学的娱乐中心 ■	elementary school's recreation center
门房 ■	caretaker's dwelling
公共区域 ■	common premise
操场 ■	playground

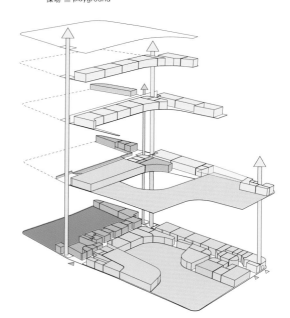

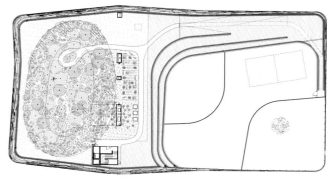

五层 fourth floor

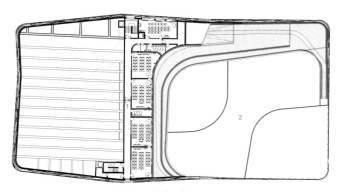

1 小学 2 操场
1. elementary school 2. playground
四层 third floor

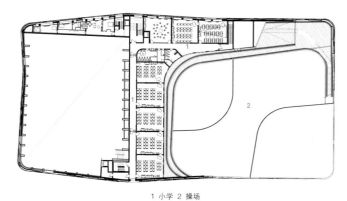

1 小学 2 操场
1. elementary school 2. playground
三层 second floor

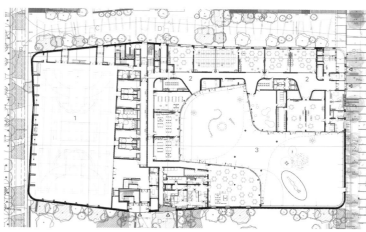

1 体育馆 2 学前班 3 操场
1. gymnasium 2. nursery school 3. playground
一层 ground floor

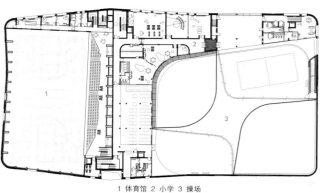

1 体育馆 2 小学 3 操场
1. gymnasium 2. elementary school 3. playground
二层 first floor

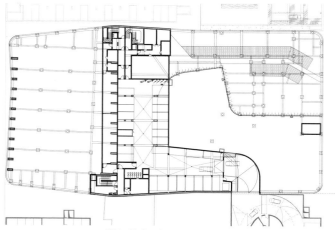

地下一层 first floor below ground

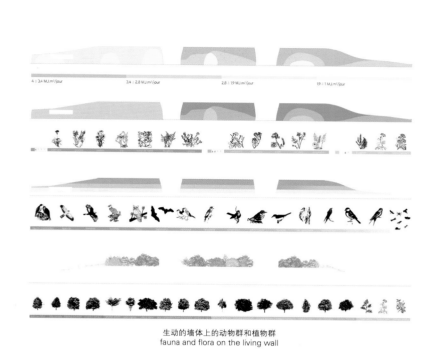

生动的墙体上的动物群和植物群
fauna and flora on the living wall

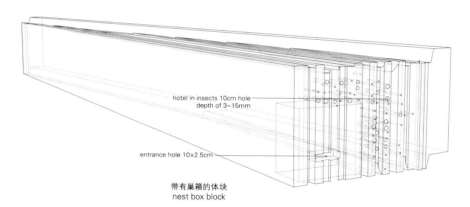

带有巢箱的体块
nest box block

view of one another (one day the preschool children will be in primary school). The entire building shelters a primitive natural environment which acts as a more or less long-term catalyst for biodiversity at the heart of the larger site. Indeed, this structure is alive in that its appearance changes. Through its function as a foundation for the landscape, it presents an envelope which will be different in five or ten years' time, as all the uncertainty of nature, which does not necessarily appear where one might expect.

The "bark" of the project, the living wall, is made of prefabricated blocks of concrete. These blocks present two different types of texture. The visible side is smooth, polished; it reflects the light. The other sides are ribbed, with a rough, rugged texture. This difference in surfaces helps to channel water towards the sides of blocks, thus avoiding trickling on the visible side and premature aging. The opposition of these two textures also emphasizes the depth of the facade and enhances its relief. On the lower section, and up to a height of around two meters, the freestanding wall is smoothed or slopes outwards, barring access to outsiders but also potential predators. The indentations of the side faces of the wall also encourage vegetation(bowls for ferns, rough concrete for mousse); small hollows and folds are aimed at animals(overhangs for swallows, porous nooks for insects) and act as an invitation to nesting for several varieties of bird.

The roof is a real hanging garden, twelve meters above the gymnasium. It is a home to three levels of vegetation: A prairie, plants planted in 50 centimeters of earth, a shrub land fringe and a woodland island planted in 1 meter of earth. The continuity of environmental corridors created by flora enables natural communication between species. This elevated garden has two functions: first, for the fauna of the wall and second for its own fauna. It is rich in resources(additional living habitats, nutrition, etc.) ensuring the success of biodiversity.

1 体院馆 2 操场 3 幼儿园 4 小学
1. gymnasium 2. playground 3. nursery school 4. elementary school
A-A' 剖面图 section A-A'

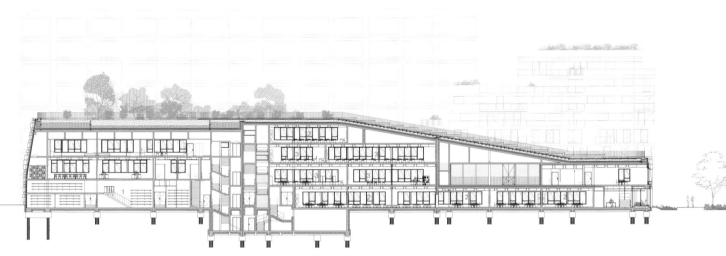

1 体育馆 2 幼儿园 3 小学
1. gymnasium 2. nursery school 3. elementary school
B-B' 剖面图 section B-B'

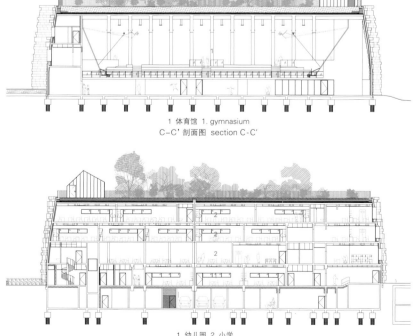

1 体育馆 1. gymnasium
C-C' 剖面图 section C-C'

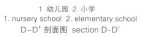

1 幼儿园 2 小学
1. nursery school 2. elementary school
D-D' 剖面图 section D-D'

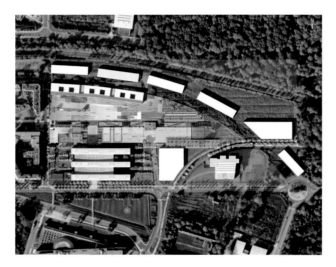

巴黎东区科学与技术区

Jean-Philippe Pargade Architecte

法国巴黎东区科学与技术区位于占地7公顷的马恩拉瓦莱,学校里建有一处比耶维涅空间,以纪念法国国立路桥学校督导和巴黎地铁之父费尔杰斯·比耶维涅。这处空间设计完善了笛卡尔的簇状城市肌理,并与新城区相结合,符合法国生态、可持续发展和能源部门出台的政策。建筑旨在整合政府在马恩拉瓦莱设立的培训和研究设施,从而建立一个注重可持续发展的城市中心。

法国巴黎东区科学与技术区集高等教育、训练、研究和工程等各学科为一体,是一个重点创新项目,旨在采用全新的方式去设计、建造、发展和管理城市。

比耶维涅空间占地将近40 000m²,其中包括25 000m²的办公区和10 000m²的实验室(包括化学、光学和材料实验室)和一个测试板(50m×10m),一个会议中心(会堂可容纳250人,此外还有会议室)以

及一间可容纳1700人同时就餐的食堂。

该园区致力于维持首都中心区域的地理平衡,距巴黎市中心20公里,旨在成为巴黎东区域发展的决定要素。在笛卡尔簇状建筑中心,比耶维涅建筑是一个各类研究机构可以协同合作的地方。

从更宏观的层面看,由于大巴黎计划的倡议(全新的巴黎全球化计划),该项目将成为国内外主要的绿色技术研究中心,同时也将成为世界最大的研究城市问题的中心。

该项目的设计理念是以设计出不拘一格的城市建筑(即笛卡尔园区)的雄心为基础的。它在街区中心展现了一片广阔的、在一座新的城镇内独树一帜的公共空间,强化了城市结构并且连接了已有的设施。在延伸的绿色屋顶的背景下,起伏的长形混凝土结构被一座景观花园覆盖,与场地平坦的平面形成鲜明的对比。这个长达200m的绿色嵌板形成了一个大型的中心花园,花园在园区和城市中央不断扩大。建筑北侧是毗连的密集楼群,另一侧是大型公园释放的清新空气,两相对比,进一步突出了城市中的自然景观。

建筑设计将一切利于可持续发展的创新技术都结合起来,突出环境、能源管理、服务与维护、温湿度和视觉舒适度等要素之间的关系。

建筑设计尤其强调实现能源低消耗的目标。若应用整体生态气候学的设计,这一目标便有可能实现,设计要素包括朝向和建筑保温(大部分的南立面面向园区,吸收太阳能,北立面相对封闭,墙体采用高热惯量的保温材料实现保温效果)、自然通风、雨水收集、保温材料。建筑制热和制冷的能源主要来源于地热能和地下水。建筑是从这种堪称完美的内涵(即进行城市可持续发展的创新研究实验室)与外在(建筑本身要具有高度环保、高效利用能源和生态节能的意识)搭配设计中汲取灵感的。

北立面 north elevation

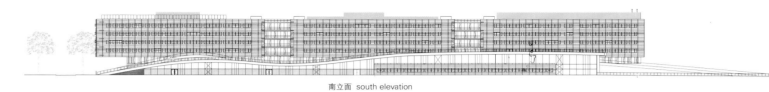
南立面 south elevation

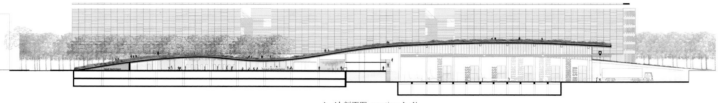
A-A' 剖面图 section A-A'

B-B' 剖面图 section B-B'

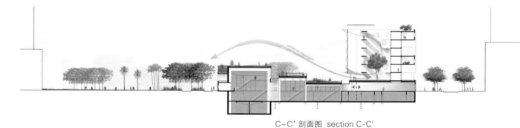
C-C' 剖面图 section C-C'

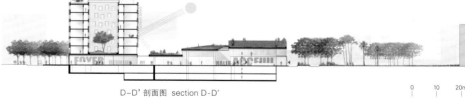
D-D' 剖面图 section D-D'

0　10　20m

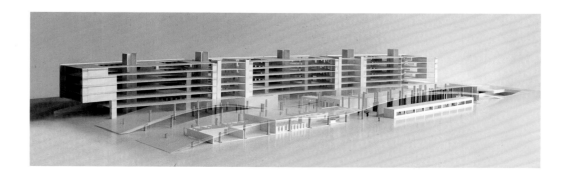

East Paris Scientific and Technical Pole

On a site covering nearly 7 hectares in Marne-la-Vallée lies the East Paris Scientific and Technical Pole which has baptised the Espace Bienvenüe in memory of Fulgance Bienvenüe, inspector general of the Ponts et Chaussées engineering school and father of the Parisian metro system. The Cité Descartes is completed by the space that has been incorporated into this new town and meets the needs of the programme imposed by the French Ministry of Ecology, Sustainable Development and Energy. Its aim is to group together the Ministry's training and research facilities on the Marne-la-Vallée site and create a center of excellence focussed on the sustainable city.

The East Paris Scientific and Technical Pole, which brings together the various disciplines of higher education, training, research and engineering, is a major innovative project intended to encourage the emergence of a new way to design, build, develop and manage cities.

Bienvenüe has a surface area of nearly 40,000m², of which 25,000m² of offices and 10,000m² of laboratories (chemical, optical and materials), as well as a test slab (50 x 10 m), a conference center (auditorium seating 250 people and meeting rooms), and a restaurant seating 1,700 People.

The Pole contributes to the major geographic balances of the capital region. 20km far from the centre of Paris, it aims at becoming a defining element in the development of the eastern Paris region. At the heart of the Descartes cluster, the Bienvenüe building is a place of specific synergies between several research institutions. On a larger scale, thanks to the Greater Paris Initiative (a new global plan for the Paris metropolitan region), the project will develop a major research pole focused on green technologies at the national and international level. It is slated to become one of the world's largest centers that addresses the issue of the urban city.

The architectural concept is based on the ambition of creating an exceptional urban event: the Cité Descartes campus. It showcases a vast public area in the center of a block unique to the new town,

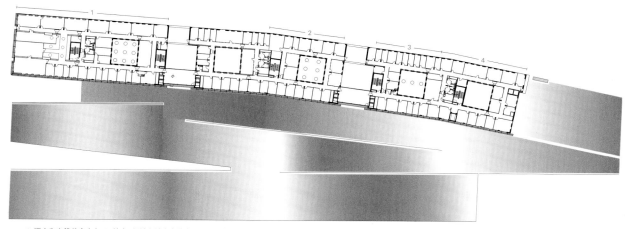

1 研究和高等教育中心 2 技术、领域和社会实验室——研究中心 3 技术、领域和社会实验室——管理办公室 4 土壤和岩石力学以及工程地质学实验室
1. research and higher education center 2. technologies, territories and society laboratory–research center
3. technologies, territories and society laboratory–management office 4. soil and rock mechanics and engineering geology laboratory
二层 first floor

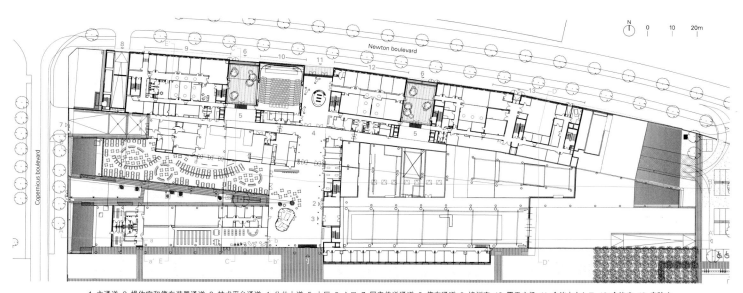

1 主通道 2 操作室和停车装置通道 3 技术平台通道 4 公共大道 5 大厅 6 入口 7 厨房传送通道 8 停车通道 9 培训室 10 露天广场 11 会议中心入口 12 会议室 13 实验室
1. main access 2. practical work rooms and parking facilities access 3. technical platform access 4. public street 5. hall 6. entrance 7. kitchen delivery access
8. parking access 9. training room 10. amphitheater 11. conference center entrance 12. meeting room 13. laboratory
一层 ground floor

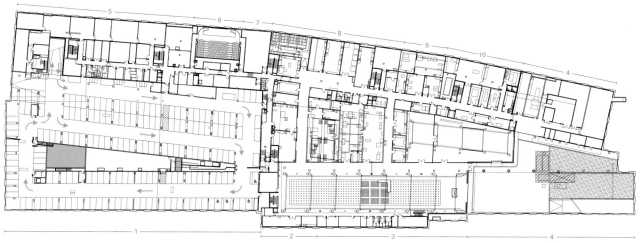

1 停车场 2 车间 3 测试平台 4 技术场地 5 档案室 6 露天广场 7 信息台 8 持久性和流变学实验室
9 土壤和岩石力学以及工程地质学实验室 10 计量和仪器单元室
1. parking 2. workshops 3. test platform 4. technical court 5. archives 6. amphitheater 7. information 8. durability and rheology laboratory
9. soil and rock mechanics and engineering geology laboratory 10. metrology and instrumentation unit room
地下一层 first floor below ground

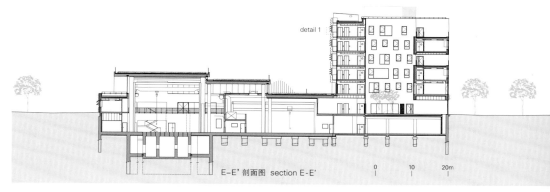

E-E' 剖面图 section E-E'

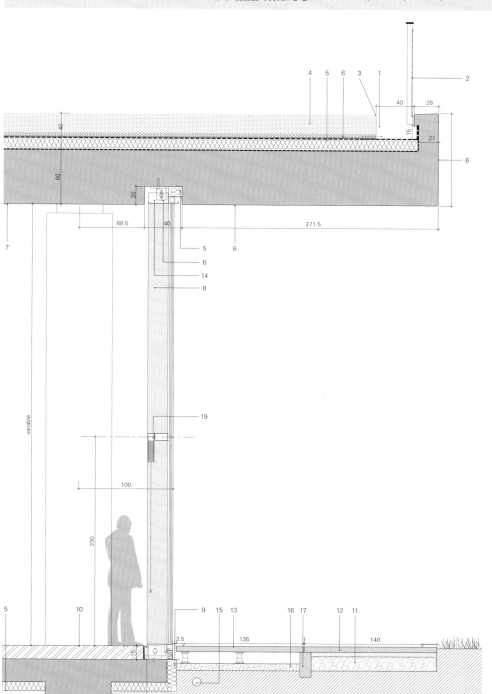

1. barren margin, waterproofing
2. balustrade hardware
3. edging, galvanized steel profile
4. green complex or green flat roofing, 26 to 30cm thick
5. insulation
6. waterproof concrete
7. untreated concrete slab
8. glazed curtain wall
9. insulation + waterproofing for buried section
10. heating floor, 15cm thick
11. gravel-cement mixture on rolled fill
12. concrete precast slab, bedded installation, thick 6cm
13. concrete precast slab on mounting, 140×40cm thick
14. box for hanging on curtain wall
15. buried peripheral drain
16. shape mortar
17. concrete edging
18. concrete refill
19. motorised blind on cross beam

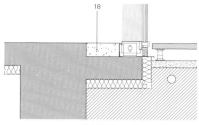

a-a' 剖面图 section a-a'

b-b' 剖面图 section b-b'

which reinforces urban structures and interlinks existing facilities. In an extension of the green setting of the campus, the creation of a long, undulating concrete structure covered with a landscaped garden contrasts with the flatness of the site. This 200-meter long green slab makes up the large central park, which is magnified at the level of the campus and the city. The contrast between the density of the buildings that abuts the block to the north and the fresh air of this vast garden enhances the urban landscape.

The architectural approach integrates all the innovations that promote sustainable development. It highlights the relationship between the environment, energy management, servicing and maintenance, hygrothermic and visual comfort.

Special emphasis has been placed on the low-energy target. This objective will be achieved through an overall bioclimatic design: orientation and insulation of the building (south facade largely opened towards the campus and retrieving solar energy/north facade more closed and thus offering an insulating wall with a strong thermal inertia), natural ventilation, rainwater collection, insulating materials. The main source of energy for heating and cooling the building is geothermal energy obtained from the groundwater body. The project thus draws its inspiration from a perfect match between a content – laboratories conducting the most innovative research in terms of sustainable development – and its container – the architecture which is highly environmental, energetic and ecologically efficient.

项目名称：East Paris Scientific and Technical Pole
地点：Cité Descartes, Champs-sur-Marne, 77420 Marne-la-Vallée, France
项目管理：Jean-Philippe Pargade
委托建筑师：David Besson-Girard
项目团队：partner_Caroline Rigaldiès / construction site director_Jean-Pierre Lamache / studies leader_ Birgit Eistert/ Christophe Aubergeon, Marco Carvalho, Joana César, Paolo Correia, Jean-Patrick Degrave, Emmanuèle Fiquet, Emilie Guyot, Antoine Hermanowicz, Joon-Ho Lee, Aline Marthon, Natacha Nass, Lucy Niney, Audrey Oster, Maxime Parin, Anne-Sophie Richard, Samuel Rimbault, Vincent Sengel, Ji Yeon Song, Marie Suvéran, Arthur Tanner, Van Hai Vu
结构工程师：Leon Grossew
基础工程师：Spie Fondations
室外木工：Rinaldi Structural / 室内木工：Bonnardel
防水、景观屋顶设计：Soprema
电气工程师：Insmatel
景观建筑师：SNC Lavalin
工程公司：Voxoa
甲方：Ministry of Ecology, Sustainable Development and Energy – Property and Housing Action Delegation
功能：research units, test laboratories, offices, lecture rooms, library, conference center, restaurant, sports facilities
面积：35,300m²
造价：EUR 95,000,000
设计时间：2008 / 施工时间：2010 / 竣工时间：2014
摄影师：
©Sergio Grazia (courtesy of the architect) - p.22~23, p.25, p.26, p.28, p.32^{bottom}, p.33, p.35, p.48~49
©Luc Boegly (courtesy of the architect) - p.30~31, p.32^{top}, p.34

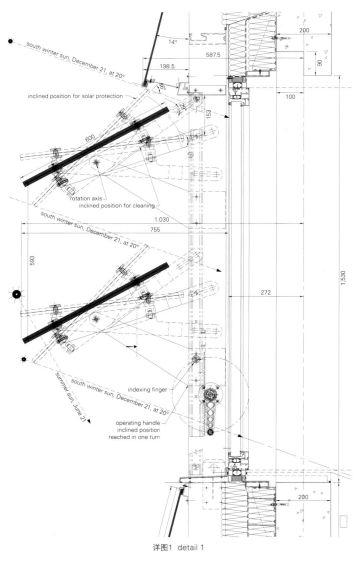

详图1 detail 1

植被25建筑
Luciano Pia

灰色建筑中的绿色自然：混合型建筑模式 Green in Grey: architecture in hybrid mode

这是一座坐落在都灵的新建筑，尽管建筑框架采用了钢筋，但是看起来更像是一片森林，树木种植在形状不规则的露台之上，覆顶的幽深小径和茂盛花园穿过池塘。整座建筑就像一片生机盎然的大森林，树上的房子就像儿时梦里面建造的树屋一样。

针对居民住宅缺乏自然的均衡性和多样性，十分必要在此街区建造一座建筑来填补这一问题。项目的目标是利用一个连续的立面来围合街区的边界，又要实现内部与街道之间的过滤功能。设计初衷是要创造一种流畅的、受人欢迎的过渡空间，从而弱化室内通往室外的通道的形象。这处流畅的、可变换的过渡空间通过指定的植物和建筑材料得以突出，从而打造出一个结构紧凑且与众不同的、透明且富于变幻的、令人身心愉悦的结构。

这是一座创意建筑，因为它是一个鲜活的机体，它会成长，会呼吸，并且因为150棵树干覆盖了露台的树木会发生变化。它们与花园里的50棵大树共同产生氧气，吸收二氧化碳，减少空气污染，降低噪音，伴随四季周而复始，每天都在变化，形成建筑内完美的微气候，冬暖夏凉，气候宜人。夏天，面向街道一侧的树木过滤阳光；冬天，阳光透过树枝的间隙洒进入房间。这里，落叶松护壁板具有柔软且充满生气的表面。金属结构如同树木一样，从一层中"生长"出来，直至屋顶，同时支撑着露台的木板；它们与植被缠绕在一起，形成了独特的立面。

项目的设计目的之一是为了提高能源利用率，因此，项目整合了多种不同的方案，如利用连续的保温外墙、防晒系统，利用带有热泵的地热能源来维持供暖和制冷系统，回收雨水来灌溉植物。

整座建筑有63间公寓，每间公寓风格各异，并配有环绕树木而建的、宽阔的、不规则形状的露台。最上面一层覆盖了种满植物的私人屋顶。

建筑内的绿色植物也是多种多样的：大盆栽可以种在露台上，或是在庭院花园里，植物墙和屋顶花园位于顶楼前侧，盆栽内可以栽种灌木或乔木，还有落叶木，高度在5.5m~8m不等，落叶植物使冬天的阳光照进房间里。绿植种类的选择根据设计不同的需求具有多样性，保证了叶子多种多样，异彩纷呈，缤纷四溢。

待枝繁叶茂时，整座建筑给人一种居住在树屋的感觉，你会感觉这便是梦中佳所。不，或许，你已然住进了美梦里。

25 Green

A new house has put down its root in Turin. Its structure is in steel and it looks like a forest where trees are rooted in terraces with irregular shapes, and ponds are crossed by footings and lush gardens covering the roofs. The building has been thought as a living forest, a house on the trees like the houses children dream of and sometimes build.

The project comes from the necessity of making a residential building to complement a block featured by lack of homogeneity and heterogeneous prospects. The aim of the project is both the construction of the block perimeter with a continuous facade and the making of a filter between the internal inhabited space and the streets. The project wants to create a flowing and smooth transition space to soften the passage from the inside to the outside where the space is always enjoyable. The smooth and changeable transition is emphasized by a targeted use of the green and the building materials so to create a structure which is compact and distinct but also transparent, mutable and enjoyable.

It is a special building because it is alive: it grows up, it breaths and it changes since 150 trees with tall trunks covering its terraces.

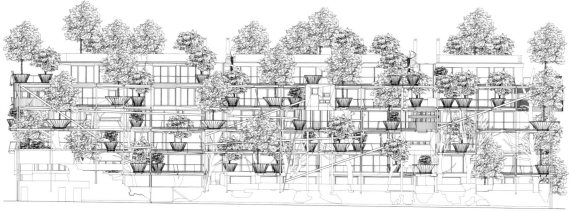

南立面 south elevation

西立面 west elevation

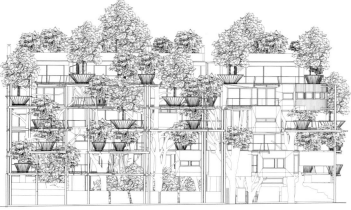

东立面 east elevation

项目名称：25 Green
地点：Via Chiabrera 25, Torino, Italy
建筑师：Luciano Pia
结构工程师：Ing. Giovanni Vercelli / 能源工程师：Ing. Andrea Cagni
绿化：gruppo corazza, maina costruzioni, de-ga s.p.a.
执行人员：DE-GA S.p.A.
照明工程师：Luce per / 墙面板设计工程师：Sindele
木结构工程师：Tesio / 考顿钢结构工程师：CCM
景观建筑师：Vivai Reviplant & Bonifacino Angelo
用地面积：3,570m² / 有效楼层面积：7,500m² / 露台面积：4,000m²
庭院花园面积：1,500m² / 景观屋顶面积：1,200m² / 盆栽面积：1,200m² / 水池面积：150m²
施工时间：2007~2012
摄影师：©Beppe Giardino (courtesy of the architect)

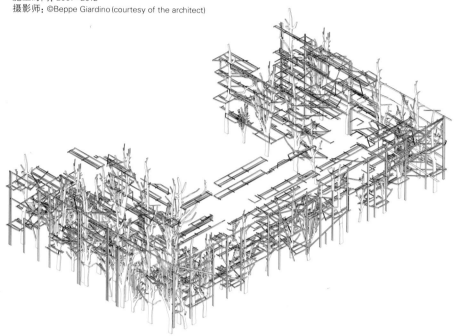

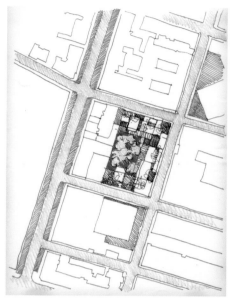

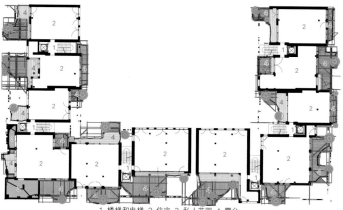

1 楼梯和电梯 2 住宅 3 私人花园 4 露台
1. stair and lift 2. residential 3. private garden 4. terrace
三层 second floor

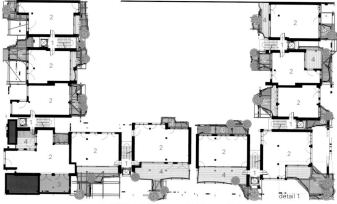

1 楼梯和电梯 2 住宅 3 私人花园 4 露台
1. stair and lift 2. residential 3. private garden 4. terrace
二层 first floor

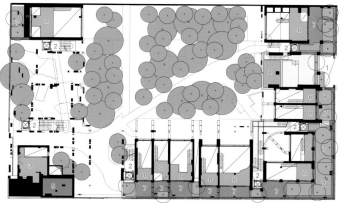

1 停车场入口 2 楼梯和电梯 3 门房 4 覆顶的花园 5 上层花园 6 住宅 7 私人花园
1. car parking entrance 2. stair and lift 3. gatehouse 4. covered garden
5. upper garden 6. residential 7. private garden
夹层 mezzanine floor

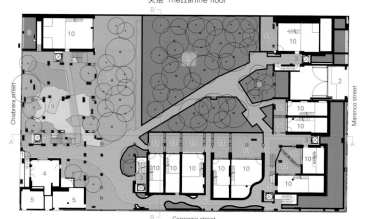

1 人行道入口 2 停车场入口 3 楼梯和电梯 4 门房 5 技术室 6 走廊 7 平地花园 8 上层花园
9 水池 10 住宅 11 私人花园
1. pedestrian entrance 2. parking entrance 3. stair and lift 4. gatehouse 5. technical room
6. gallery 7. flat garden 8. upper garden 9. water area 10. residential 11. private garden
一层 ground floor

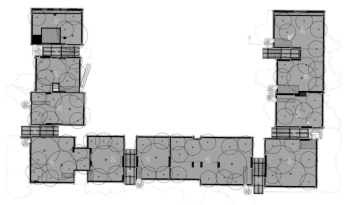

1 楼梯和电梯 2 屋顶花园
1. stair and lift 2. roof garden
屋顶 roof

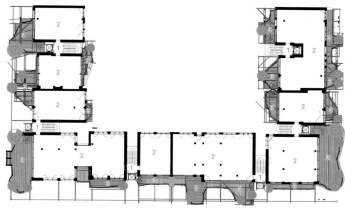

1 楼梯和电梯 2 住宅 3 露台
1. stair and lift 2. residential 3. terrace
六层 fifth floor

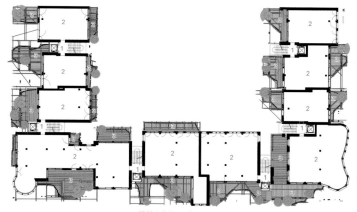

1 楼梯和电梯 2 住宅 3 露台
1. stair and lift 2. residential 3. terrace
五层 fourth floor

1 楼梯和电梯 2 住宅 3 露台
1. stair and lift 2. residential 3. terrace
四层 third floor

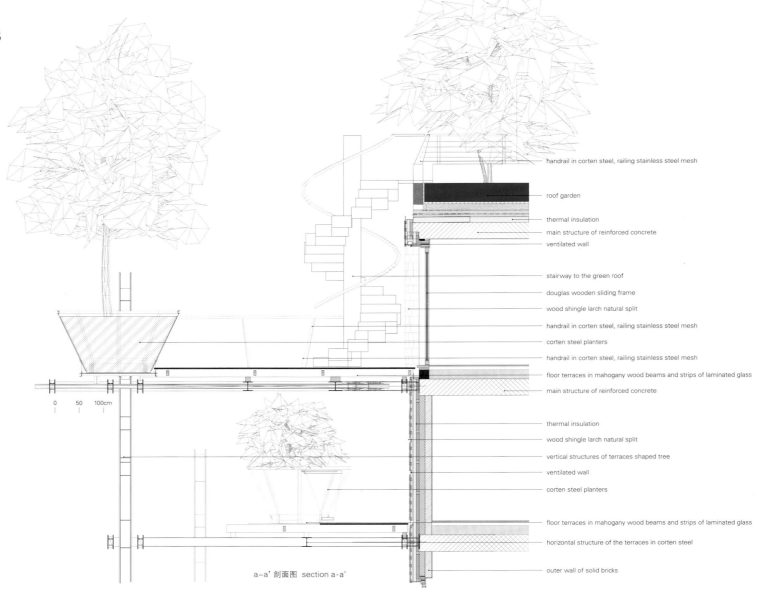

a-a' 剖面图 section a-a'

- handrail in corten steel, railing stainless steel mesh
- roof garden
- thermal insulation
- main structure of reinforced concrete
- ventilated wall
- stairway to the green roof
- douglas wooden sliding frame
- wood shingle larch natural split
- handrail in corten steel, railing stainless steel mesh
- corten steel planters
- handrail in corten steel, railing stainless steel mesh
- floor terraces in mahogany wood beams and strips of laminated glass
- main structure of reinforced concrete
- thermal insulation
- wood shingle larch natural split
- vertical structures of terraces shaped tree
- ventilated wall
- corten steel planters
- floor terraces in mahogany wood beams and strips of laminated glass
- horizontal structure of the terraces in corten steel
- outer wall of solid bricks

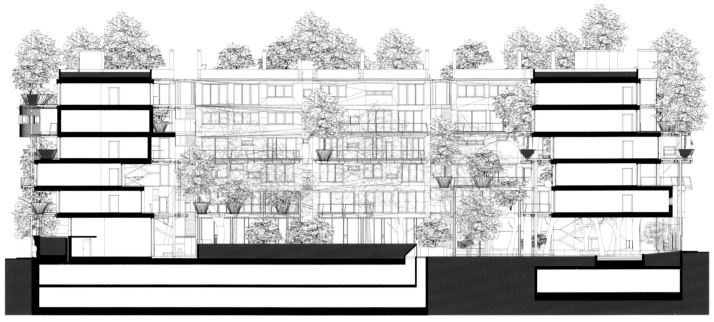

A-A' 剖面图 section A-A'

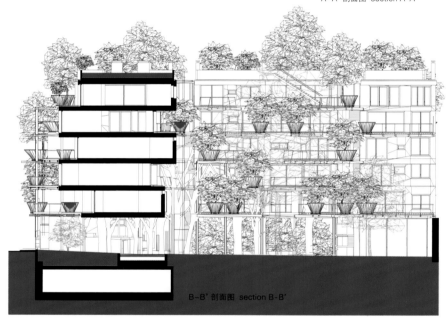

B-B' 剖面图 section B-B'

Together with 50 trees planted in the court garden they produce oxygen, absorb carbonic anhydride, cut down air pollution, protect from noise, follow the natural cycle of seasons, grow up day after day and create a perfect micro-climate inside the building so diminishing the fall and rise in temperature in summertime and wintertime.

The streets side solid wood filters the sunlight in summer, while in winter lets the light break into the house. The wainscot in larch shingles has a sort of soft and vibrant surface. The metal structures look like trees and they "grow" from the ground floor to the roof while holding up the wooden planking of the terraces: they become entwined with the vegetation to form a unique facade.

One of the aims of the project is the increase of the energetic efficiency and for this reason several integrated solutions have been adopted: continuous insulation, sun protection, heating and cooling systems which make use of the geothermal energy with heat pumps and recycling of the falling rain to water the green.

There are 63 residential units in the building and they are all different and fitted with wide terraces of irregular shapes that surround the trees. The last floor is covered with private green roofs.

The green is diversified: big vases on the terraces and court gardens; green walls and roof gardens just in front of the lofts. In the vases there are trees or shrubs of different heights from 2.5m to 8m. Deciduous species have been planted to have sun irradiation in winter too. The choice of the species, diversified according to the different needs, has been made to grant a variety of leaves, colors and flowering.

When all the green is fully blooming it gives the feeling of living in a tree house. You can dream of a house or live in a dream.

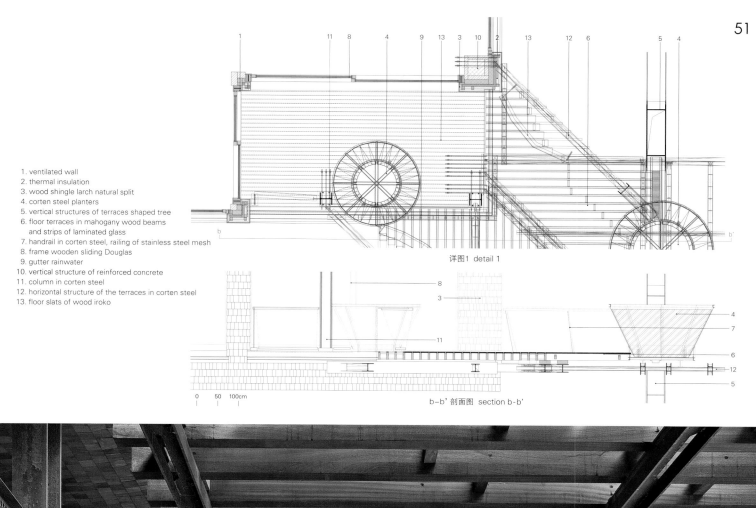

1. ventilated wall
2. thermal insulation
3. wood shingle larch natural split
4. corten steel planters
5. vertical structures of terraces shaped tree
6. floor terraces in mahogany wood beams and strips of laminated glass
7. handrail in corten steel, railing of stainless steel mesh
8. frame wooden sliding Douglas
9. gutter rainwater
10. vertical structure of reinforced concrete
11. column in corten steel
12. horizontal structure of the terraces in corten steel
13. floor slats of wood iroko

详图1 detail 1

b-b' 剖面图 section b-b'

绿化改造

Vo Trong Nghia Architects

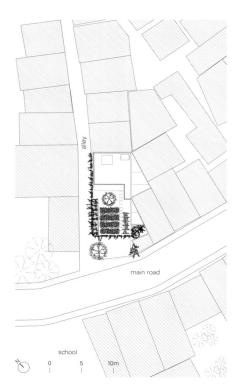

越南的迅猛发展引发了许多城市问题。植被覆盖空间越来越少，供电短缺，洪水泛滥等。与日俱增的摩托车导致严重的交通拥堵和空气污染。为改善城市环境恶化的现状，"绿色瀑布改造"建筑堪称是独栋建筑设计的原型，还城市一个绿色家园，为当地居民和社区提供一处既舒适又生机勃勃的居住环境。河内市中心有一幢十五年的老房子，位于一条西南向街道和西北向小巷交叉的街角，就像河内其他的老房子一样，常年处于昏暗、潮湿和发霉的环境中。因为要保证楼内的安全和封闭性，老房子使用了防护栏和百叶窗，这样的设计导致阳台一直处于闲置状态，不能发挥作用。为了改变这一状况，改造项目引入了大量的绿色植物和充足的阳光，原有的厚重混凝土楼梯被换成纤长的钢骨楼梯，形成一个三角形的采光井，可以透过从天窗射来的自然光，且保持良好的通风。采光井的墙面材料选用粗糙的大理石，可以反射和扩散自然光线。二层引入一处上空空间，以连接餐厅（一层）和书房（二层），这样可以鼓励居民进行交流，同时阳光也会透过植物立面深入到房间里。河内市在换季时，房间通常阴冷潮湿，因此一层被抬高，下面安装了通风隔层，防止潮气和湿气侵入。

改造后的房子的亮点是被称为"绿色瀑布"的绿植立面设计，不管从哪个角度看，从里面还是从外面看，这个"瀑布"都赏心悦目，令人心情愉快。老旧的防护栏拆下来，取而代之的是阳台安装了镀锌的钢框架，上面攀爬着藤蔓植物。这样，原本没用的阳台就变成了一座植物园，既保护了住户的隐私，又保证了安全。居住其中的每一个房间的人们都能欣赏到生机盎然的景色，呼吸到新鲜的空气。绿色立面的设计不仅使住户受益，也让邻居欣赏到街道一侧枝繁叶茂的怡人景象。

绿色立面和屋顶花园共同实现了降低能源消耗的目标，绿植保护建筑免受强烈的西侧射来的光线的侵害。屋顶花园种植着种类繁多的花草树木和蔬菜，幸亏有了这些植物的陪伴，当地住户意识到了保护环境的重要性，开始减少不必要的能源消耗。

这种绿植立面和屋顶系统成为可以应用到其他热带气候地区的建筑原型，实现热带国家建设绿色城市的梦想，以解决严重的城市问题。这种设计还可以减少洪水的侵害，这里的土壤层就如同蓄水池，通过增加植被来减轻城市热岛效应，而当地的居民还可以享受到绿化景观及其果实，从而促动社会交流。人们希望这个项目能够成为热带城市绿化的代表作，绿色健康家园的建立不仅能给当地居民，还会给整个城市都带来积极和广泛的影响。

Green Renovation

Vietnam's fast development raises many urban problems: less green space, electricity shortages and flooding. The increasing number of motorbikes is causing traffic jams and serious air pollution. "Greenfall renovation", a prototypical single house renovation, was designed against this backdrop, returning greenery to the city and encouraging a comfortable living environment to both the residents and neighbors. A 15 year old existing house, located in the city center, at a corner of a street on south-west and an alley on north-west, had suffered from dark, wet and moldy environment, which is a typical condition for many of the older houses in Hanoi. The house had been secured and closed by security bars and shutters, making balconies unused space. To remedy this situation, the house is renovated to live with green and abundant light. The existing massive concrete staircase was replaced with slender steel stairs, creating a triangle light well, through which natural light and ventilation are introduced from skylight. Walls of this light well are finished by lumpy marble stone to reflect

and diffuse the light. Another void was cut into first floor to connect the dining space(GF) with a study(1F), which encourages communication of residents and brings sunlight through the green facade deep into the house. The ground floor was raised to install air ventilation layer beneath, to prevent damp and condensation, a frequent problem in Hanoian houses at the turn of seasons.

The house is characterized by green facade named "Greenfall", a pleasant green waterfall which attracts both from interior and from exterior. Old security fences were removed and replaced with galvanized steel trellis, attached to existing balcony, on which climbers grow. So the unused balconies are transformed into spaces for greens, keeping the privacy and security for the residents. From the interior, every room can enjoy the view of greenery and get fresh air through it. This green facade is not only for residents but also for neighbors, providing attractive leafy-scape to the streets.

The green facade and the roof garden function together to reduce energy consumption: they protect the building from harsh west sunlight. Many kinds of vegetables and flowers are planted on the green roof, as well as a tree. Thanks to the plenty of greenery, the residents became more aware of environment and started to take actions to reduce unnecessary energy.

This system of green facade and roof is prototypical and applicable to all the buildings in tropical climates. It can realize a potential Green City in tropical countries, offering solutions for their serious urban problems. It can reduce the risks of flooding, where the layer of soil functions as retention basins, reducing the urban heat island effect through the increase greenery, while the building promotes social interaction through sharing the green-scape and its food products. It is hoped that this project will be a modest model to greening tropical cities, where benefits of a healthy green home can be shared by not just the occupants but have a wider positive impact on the city.

项目名称：Green Renovation
地点：Hanoi, Vietnam
建筑师：Vo Trong Nghia Architects
主要建筑师：Vo Trong Nghia, Takashi Niwa, Tran Thi Hang
项目建筑师：Ngo Thuy Duong, An Viet Dung
承包商：Wind and Water House JSC.
用地面积：110.5m²
有效楼层面积：387.9m²
竣工时间：2013.2
摄影师：
Courtesy of the architect - p.54, p.55, p.56
©Hiroyuki Oki (courtesy of the architect) - p.57, p.58, p.59

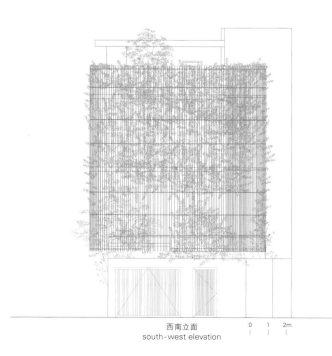

西南立面
south-west elevation

1 停车场 2 起居室 3 餐厅 4 厨房 5 卧室 6 礼拜室 7 储藏室 8 屋顶花园
1. parking 2. living 3. dining 4. kitchen 5. bedroom 6. worship room 7. storage 8. roof garden
A-A' 剖面图 section A-A'

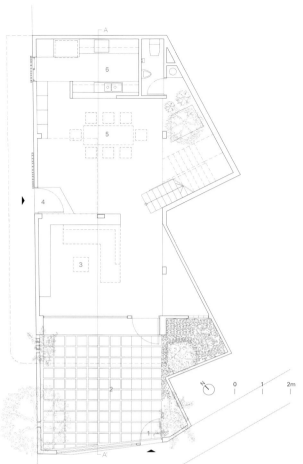

1 入口 2 停车场 3 起居室 4 次入口 5 餐厅 6 厨房
1. entrance 2. parking 3. living 4. sub entrance 5. dining 6. kitchen
一层 ground floor

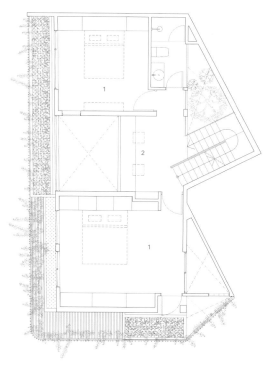
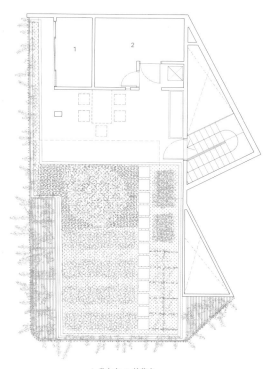

1 卧室 2 书房
1. bedroom 2. study room
二层 first floor

1 发电室 2 储藏室
1. generator room 2. storage
四层 third floor

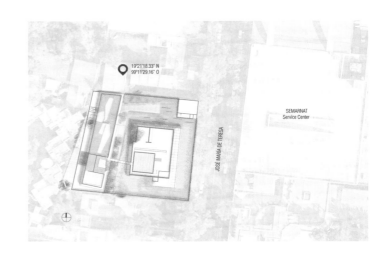

猎鹰总部二期

Rojkind Arquitectos + Gabriela Etchegaray

在猎鹰总部一期于2003年竣工后,甲方仍然委托Rojkind建筑事务所对其进行扩建。猎鹰公司是一家提供医疗设备和器械的公司,随着机构不断地发展壮大,其对办公空间、展销厅和员工生活区的需求也在增加。

猎鹰总部坐落于墨西哥城的南部St Ángel附近的一个传统居民区中心,这一地区保留着殖民时期的特色建筑、令人叹为观止的公园、低矮的小楼和鹅卵石街道,以一座古城闻名内外。

位于现在居民区的猎鹰一期的主要建筑被翻新成一个漂浮在花园里的黄色玻璃盒子结构,立面后方的空间进行了景观美化,并且战略性地形成一个可以看到户外景色的框架。猎鹰二期的设计理念是对花园进行延伸,使其与主体建筑相得益彰,以便保护园林景观视野,强调内部和外部之间的视觉联系。

猎鹰二期项目就在一期黄色玻璃盒子结构的后身,其场地现有的建筑被拆除,为新地下停车场和两层的建筑结构留出空间。建筑设计所面临的巨大挑战之一是与原有建筑的距离太近。Gabriela Etchegaray受Rojkind建筑事务所的邀请一起参与这个二期项目,这个团队决定建造一个简约透明的矩形盒子结构作为扩建的结构,这一设计既能保证充足的自然光照射,又可以保证望向花园的视野不受阻碍。这个简约透明的矩形玻璃幕墙前置一个由510个盆栽组成的控光层,这些盆栽错落排列,使立面看上去像是花园的延续。这些滴灌式盆栽利用被动式太阳能有效降低了建筑室内的温度,通过双高的中庭,花园被拉入方形玻璃结构的内部,而中庭还赋予室内一种高耸的视觉感受。垂直流线区统一位于建筑后方,使整座建筑保持了一种透明感。

一座桥将扩建建筑的屋顶花园与猎鹰一期顶层新建的员工食堂连接起来。建筑周围大面积的自然景观以及一个新的街道入口使整体设计浑然天成,融为一体。

Falcón Headquarters 2

Following the successful completion of Falcón 1 Headquarters in 2003, the client retained Rojkind Arquitectos for the design of its expansion. Falcón, a supplier of medical equipments and instruments, had outgrown its previous space and was in need of additional office area as well as space for a showroom and employee amenities.

Falcón Headquarters are located at the heart of the traditional, and mainly residential, neighborhood of San Ángel in the southern quadrant of Mexico City. This historic area is known for its colonial architecture, amazing gardens, low buildings and cobblestone streets.

In Falcón 1, the main house in a walled-in existing residential complex was completely renovated and became a yellow glass box floating in the garden with interstitial landscaping behind its facade and strategically framed views to the outside. The new Fal-

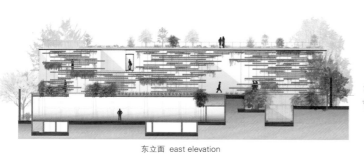

东立面 east elevation　　　　　西立面 west elevation

cón 2 expansion was conceptualized as an extension of the garden itself so that it would complement the main building and preserve its green views; visual connection between interior and exterior was emphasized.

The new building was positioned directly behind the yellow glass box of Falcón 1 on the footprint of an existing structure that was demolished to allow room for a new underground parking and two story structure. A main challenge to overcome was the proximity to the existing building. Rojkind Arquitectos invited Gabriela Etchegaray to work together in the design of this project. The team decided on a simple transparent rectangular box to contain the expansion with plenty of natural light and undisturbed garden views. A simple transparent glass curtain wall was fronted with a sun control layer composed of 510 modular planters arranged in an offset linear configuration that continued the garden feel over the facade. The drip-irrigated planters also help cool the building in a passive solar way. The garden was then strategically taken to the inside of the glass rectangle structure through a double height atrium that gave the interior a lofty feel. Vertical circulations were grouped at the back to preserve the sense of transparency.

A bridge conneced the rooftop park of the expansion with a new employee dining hall placed over the existing Falcón I building. Extensive landscape of the surrounding ground and a new street entry structure complete the overall design.

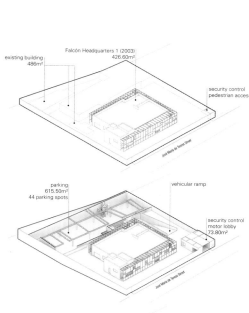

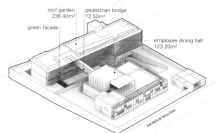

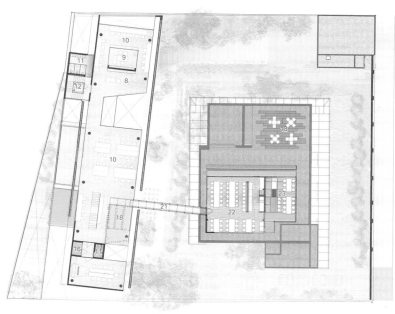

二层 first floor

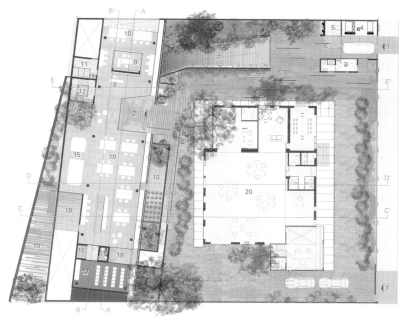

一层 ground floor

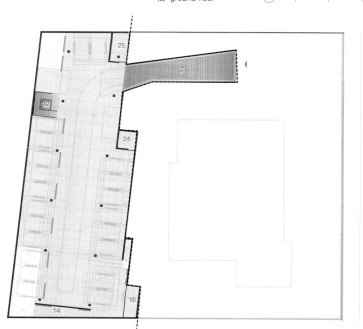

地下一层 first floor below ground

1 大堂	9 私人办公室	18 露台
2 通道猎鹰总部二期的主要通道	10 办公区	19 花园
3 安全控制室	11 卫生间	20 猎鹰总部一期
4 应急设备间	12 流线	21 人行桥
5 资源回收室	13 庭院	22 员工餐厅
6 车辆通道	14 停车场的道路	23 室外就餐区
7 车辆坡道	15 展览室	24 机械间
8 接待处	16 储存室	25 泵房
	17 放映室	
1. lobby	9. private office	18. terrace
2. main access to Falcón Headquarters 2	10. office area	19. garden
3. security control	11. restroom	20. Falcón Headquarters 1
4. emergency plant	12. circulation	21. pedestrian bridge
5. recycling room	13. courtyard	22. employee dining hall
6. vehicular access	14. parking street	23. exterior dining area
7. vehicular ramp	15. showroom	24. mechanical room
8. reception area	16. storage	25. pump room
	17. screening room	

项目名称：Falcon Headquarters 2
地点：Mexico City, Mexico
建筑师：Michel Rojkind, Gerardo Salinas, Gabriela Etchegaray
项目团队：Barbara Trujillo, Gerardo Villanueva, Gerardo Reyes, Carlos Campos, Andrea León, Rosalba Rojas, Media, Lorena García, Cordero Sasía
结构工程师：MONCAD BSI Consultores
MEP顾问：2M / 景观顾问：Ambiente Arquitectos
用地面积：2,212m² / 总建筑面积：2,090m²
设计时间：2012 / 施工时间：2012-2014
摄影师：©Jaime Navarro (courtesy of the architect)

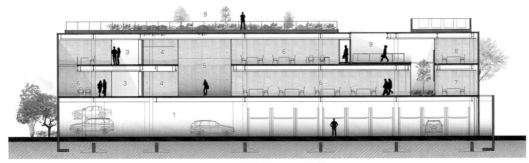

A-A' 剖面图 section A-A'

1 停车场	1. parking
2 露台	2. terrace
3 私人办公室	3. private office
4 接待处	4. reception area
5 入口大厅	5. entry hall
6 办公区	6. office area
7 放映室	7. screening room
8 会议室	8. meeting room
9 屋顶花园	9. roof garden

B-B' 剖面图 section B-B'

1 停车场	1. parking
2 放映室	2. screening room
3 会议室	3. meeting room
4 楼梯	4. circulation stair
5 办公区	5. office area
6 私人办公室	6. private office
7 屋顶花园	7. roof garden

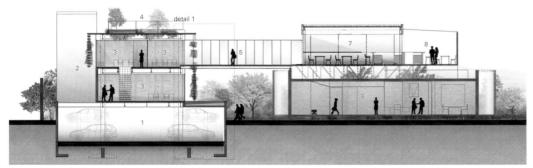

C-C' 剖面图 section C-C'

1 停车场	1. parking
2 露台	2. terrace
3 办公区	3. office area
4 屋顶花园	4. roof garden
5 人行桥	5. pedestrian bridge
6 猎鹰总部一期	6. Falcón Headquarters 1
7 员工餐厅	7. employee dining hall
8 室外就餐区	8. exterior dining area

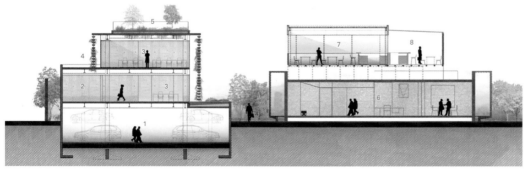

D-D' 剖面图 section D-D'

1 停车场	1. parking
2 展览室	2. showroom
3 办公区	3. office area
4 露台	4. terrace
5 屋顶花园	5. roof garden
6 猎鹰总部一期	6. Falcón Headquarters 1
7 员工餐厅	7. employee dining hall
8 室外就餐区	8. exterior dining area

E-E' 剖面图 section E-E'

1 停车场	1. parking
2 流线	2. circulation
3 接待处	3. reception area
4 办公区	4. office area
5 屋顶花园	5. roof garden
6 安全控制室	6. security control

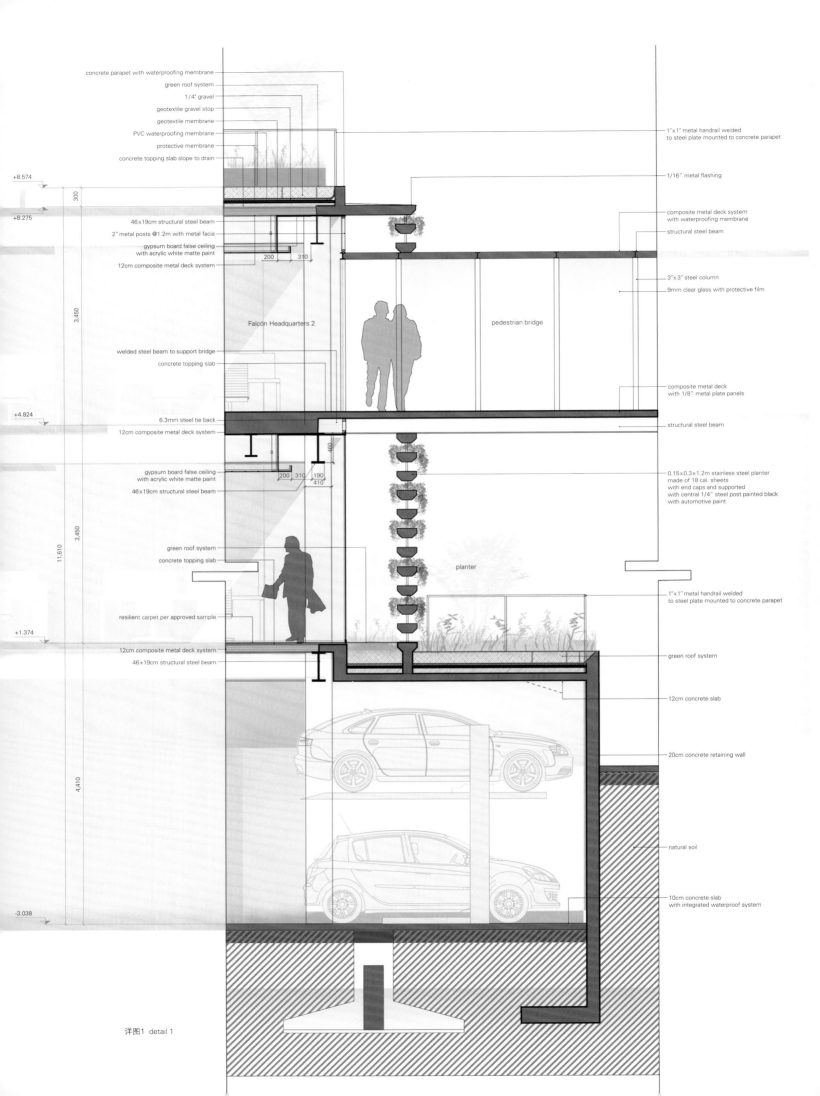

详图1 detail 1

灰色建筑中的绿色自然：混合型建筑模式 Green in Grey: architecture in hybrid mode

Point 92建筑
ZLG Design

Point 92建筑于2012年末竣工，建筑简单地以其占地面积命名，占地372 310m²，包含一座19层的塔楼，内设18 580m²的办公区。

这栋大型建筑坐落在一处斜坡场地内，俯瞰白沙罗柏兰岭及交错的街道。洋洋洒洒的阳光与建筑外立面交相呼应，加上匠心独具的建筑师们精心挑选的建筑材料，将这种氛围渲染出纯真且永恒的美感。

当访客们通过一个缓慢上升的扶梯缓缓步入大厅时，就会看到美得令人窒息的白沙罗柏兰岭的景象。尽管占地面积小，围绕场地种植的500多棵树使人仿佛置身于沿主抵达楼层种植的茂密雨林的感觉。人们可在露天平台观赏景色，在随意设置的水泥长椅上小憩片刻，顶部使用船用胶合板制造的木质灯笼与其紧紧依附的混凝土原料营造出一个具有舒缓氛围的天篷。前台接待区向内凹进，而起伏的企口墙体营造出欢迎游客的形象，并且提醒游客们注意建筑立面墙体的几何图形设计。

办公楼层平面的设计特点是留出一处由几层上空空间构成的空间，这处空间由花园和垂直设置的植物网连接。大堂中心的支柱利用不同宽度的横梁来支撑每一侧，而每一根横梁都与不同楼层的承重力相对应。这些楼层的承重力支撑起垂直的空间，也就是花园。露台被阳光照亮，其设计允许游客从花园外部欣赏风景。

为了证明简单的当地材料也可以建造出吸引人的独特建筑，建筑师对当地的材料进行了仔细的加工和精心的设计，以营造一种自然且令人愉悦的感觉。于是，白色的混凝土和当地造船用的胶合板被选为主要的建筑材料。为了解决在斜坡上施工的难题，设计选择使用现浇混凝土墙体，而非用普通的预制混凝土。仅仅使用几组金属框架结构是无法达到规划的目标的，如果立面按照常规顺序来浇注的话，还要保证墙面建造的一致性。因此，建筑师采用随机浇注的顺序。立面的倾斜墙面优雅地从斜坡上升起，沿续了场地的自然形状，末端的矩形结构为这个位于白沙罗柏兰岭的不断变化的建成环境画上完美句点。这座拥有独特外形和优雅立面的建筑成为该地区著名的地标建筑。

马兰西亚绿色建筑认证机构也将Point 92建筑认定为绿色建筑。Point 92建筑设计的主要环保特点是现浇的白色混凝土立面，150cm厚的墙面上的洞口的面积占38%，在保证最小热增量的同时，最大限度地保证了办公区的室内光线。

Point 92建筑还是一座认证的多媒体超级走廊(MSC)计算机中心建筑。

Point 92

Point 92 was completed in late 2012, and the building was simply named after the size of the site itself. It sits on a small site of 0.92 acres and consists of a single 19-story tower with 200,000 sq feet of office space.

This formidable building is situated on a sloped site, overlooking Damansara Perdana and its ribbons of roads. The play of ambience through the incorporation of day lighting with facade treatment is a true timeless aesthetic, with the impeccable play of materials the architects choose.

Visitors will get to experience a breathtaking view of Damansara Perdana as they arrive at the lobby through a slowly rising escalating step way. Despite its small size, the planting of about 500 trees around the site gives rise to the feeling of dense vegetation around the main arrival floor. This deck allows visitors to take in the view by providing randomly placed precast concrete benches that encourage respite while wooden lanterns with marine plywood ceilings buttoned tightly against raw concrete create a soothing

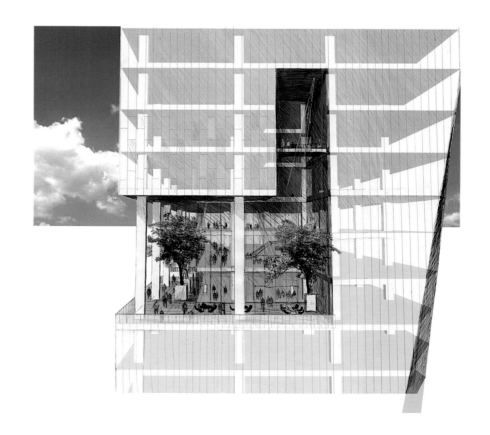

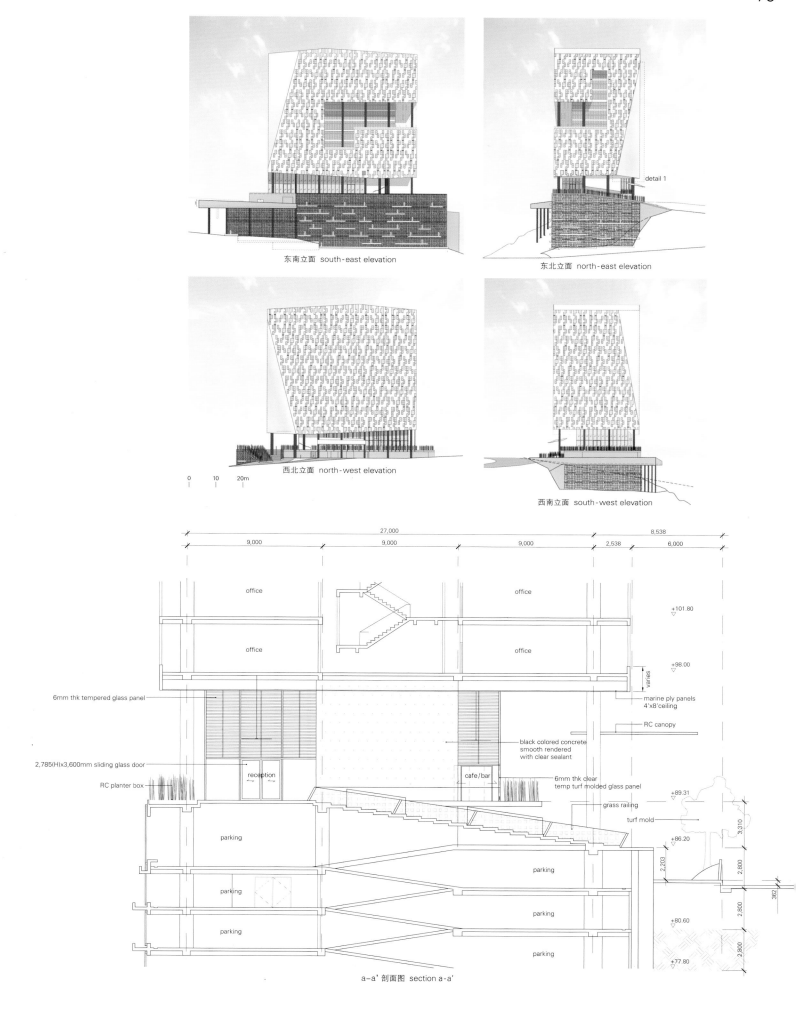

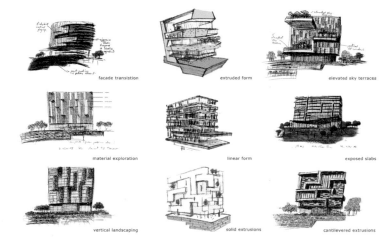

facade transition | extruded form | elevated sky terraces
material exploration | linear form | exposed slabs
vertical landscaping | solid extrusions | cantilevered extrusions

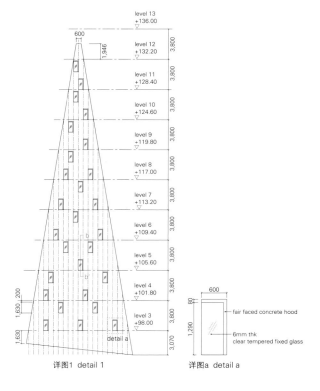

详图1 detail 1   详图a detail a

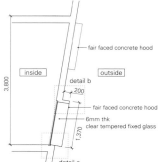

b-b' 剖面图 section b-b'

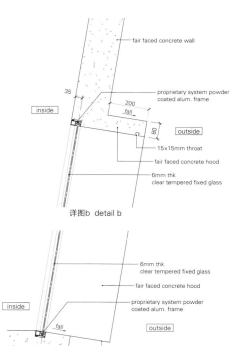

详图b detail b

详图c detail c

canopy. The reception desk recesses and the undulating rebated wall greets the visitors and reminds them of the graphics and geometry of the building's facade walls.

As an office plan, the generating feature of the floor is in fact a space which comprises of several levels of voids connected through gardens and meshes of vertical planting. The center's supporting column is braced to either side with beams of different thickness of beams each corresponding to different floor force framing the vertical space that is the garden. The terrace lit and designed to give the viewers from outside of the garden.

To show that simple local materials can be crafted to create a building that is appealing and distinctive, local materials were carefully designed and crafted to give a feel that is natural and pleasing. White concrete and local marine plywood were chosen as the main materials. To overcome the problem of building on a slope, the design opted for in situ concrete wall instead of the usual precast concrete solutions. Not only was it necessary to use metal formwork in sets to meet a target schedule, it is also informed of the uniformity issues if facade was casted in regular sequences. Hence the randomly casted sequence was archieved. Elegantly rising from the slopes, the slanting walls in the facade continue the natural geometry of the site while the rectangular punctuations complement the ever developing built environment in Damansara Perdana. The distinctive form and elegant facade have made the building a well-known landmark in the area.

Point 92 is also a certified green building by the Green Building Index (GBI) of Malaysia. A key green feature is the in situ placed white concrete facade, which comprises of 150mm thick walls with only 38% openings for windows, minimizing heat gain while maintaining optimum natural light in the office spaces.

Point 92 is also a certified MSC (Multimedia Super Corridor) Cyber Center Status Building.

项目名称：Point 92 / 地点：Selangor, Malaysia
建筑师：Huat Lim, Susanne Zeidler
Key team members_Jessica Wang, Wong Kum Tak, Willian Fong
结构工程师：JPS Consulting Engineering
机械工程师：MEP Engineering Sdn. Bhd.
规划师：PNR Consultant Sdn. Bhd.
开发商和甲方：Tujuan Gemilang Sdn. Bhd.
造价顾问：YSCA Consulting Sdn. Bhd.
主要承包商：CLOB Kumpulan CLO Bersekutu Sdn. Bhd.
竣工时间：2012
摄影师：courtesy of the architect

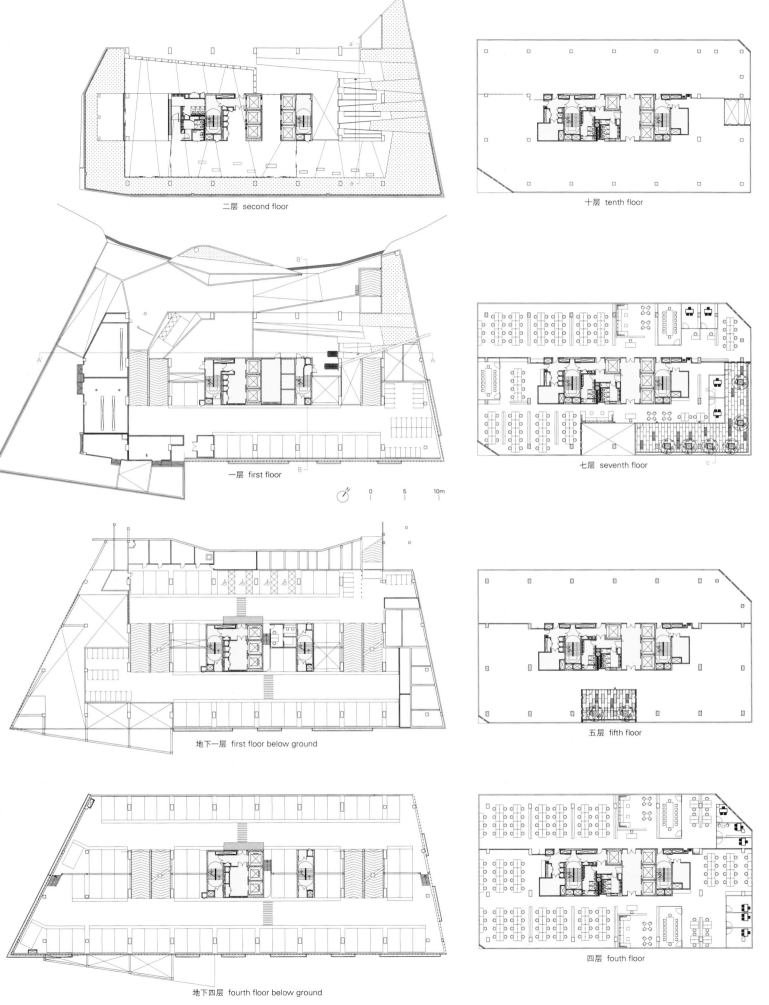

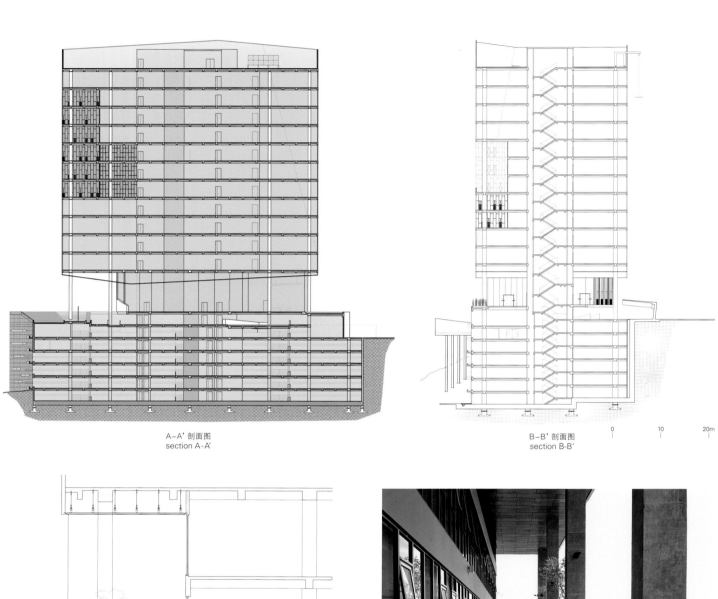

A-A' 剖面图
section A-A'

B-B' 剖面图
section B-B'

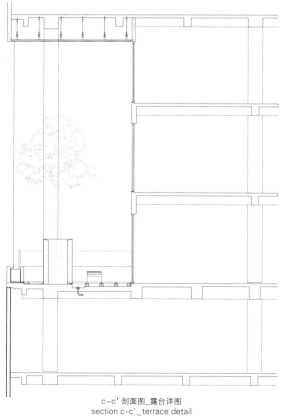

c-c' 剖面图_露台详图
section c-c'_terrace detail

垂直森林建筑
Boeri Studio

灰色建筑中的绿色自然, 混合模式建筑 / Green in Grey architecture in hybrid mode

坐落于米兰新门伊索拉区的"垂直森林"项目由意大利Hines事务所设计，是更大规模的改造项目的一部分。

米兰的垂直森林项目由两个塔楼组成，高度分别为80m和12m，建筑内种植了480棵大中型树木、300棵小型树木、11 000株常绿植物和地被植物，以及5000株灌木，在超过1500m²的城市表面种植了相当于20 000m²的森林和丛林。垂直森林的建筑理念是用树叶变化的颜色代替传统的城市表面建筑材料，以此来装饰建筑外墙。生态建筑师突破技术和机械方法的局限，依靠植物形成的屏障，来创造宜人的微气候，并且过滤自然光线，从而实现环境的可持续发展。

垂直森林项目增强了生态多样性，促进了城市生态系统的形成，在该系统内，不同的植物形成了一处独立的垂直环境，鸟类和昆虫可以栖居于此（最初估算约有1600种鸟类和蝴蝶），因此该建筑创造了各类动物和植物可再生的自然因素。

垂直森林也有助于形成微气候，过滤城市中的粉尘污染。各种植被促进微气候的形成，来增加空气湿度，吸收二氧化碳和粉尘，释放氧气，吸收辐射，减轻噪音污染等。

垂直森林还可以有效控制和减缓城市的扩张。根据城市建筑的密度计算，每一座塔楼可为独户家庭住宅和建筑提供约50 000m²的面积。

根据建筑的朝向和立面的高度而制定的物种选择和分布方案是一组植物学家和生物学家三年期的调查结果。建筑内的植物都在培育基地提前培育，使它们适应建筑阳台内相似的环境。

垂直森林是一个不断发展变化的城市地标建筑，建筑外立面的颜色会随着季节和植物属性而变化，为米兰人提供了一个永恒变换的城市景观。

花盆和植被由公寓居民共同负责管理和维护，包括所有植物的保养、替换以及花盆中的植物的数量计算等。

根据对微观气象的研究，植物的灌溉量是由气候特点决定的，立面的曝光量以及每层植被的分布都可能形成不同的灌溉需求。

Vertical Forest

The "Vertical Forest" was inaugurated in Milan in the Porta Nuova Isola area, as part of a wider renovation project led by Hines Italy. Milan's Vertical Forest consists of two towers of 80m and 112m, hosting 480 large and medium trees, 300 small trees, 11,000 perennial and covering plants and 5,000 shrubs, the equivalent – over an urban surface of 1, 500m² – of 20,000m² of forest and undergrowth. The Vertical Forest is an architectural concept which replaces traditional materials on urban surfaces using the changing polychromy of leaves for its walls. The biological architect relies on a screen of vegetation to create a suitable micro-climate and filter sunlight, and reject the narrow technological and mechanical approach to envi-

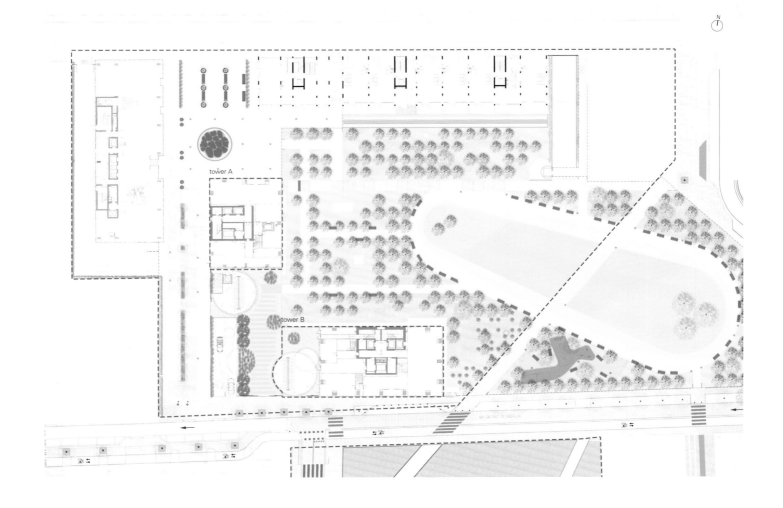

1. ventilated facade coated by slabs in stoneware of gray anthracite color with translucent effect
2. mobile clenching in aluminum cut by thermal break of gray anthracite color with double glazing in structural bonding
3. fixed clenching in aluminum cut by thermal break of gray anthracite color with double glazing in stuctural bonding
4. ventilated facade coated by slabs in printed glass according to the color of window frame
5. parapet in staves of composite material composed of aluminum slab of white pearl reflective solar
6. veil in panels of composite material composed of the slabs of aluminum of gray anthracite color with translucent effect
7. shelters in metal structure with coating in composite materials composed of the slabs of aluminum of red carmine color
8. brise-soleil in profiles of natural aluminum rectangular in section
9. continuous facade in glass with profiled mullions and transoms in aluminum of gray anthracite color
10. entrance shelter in metallic structure coated in composite material composed of the slabs of natural aluminum
11. openable clenching
12. external fence in steel bars continuing the one of the park
13. trees up to 7.5m of the species indicated in the technical report
14. trees up to 6m high indicated in the technical report
15. trees up to 2.4m high indicated in the technical report
16. bushes up to 2.4m high indicated in the technical report
17. low & drooping shrubs of the species indicated in the technical report

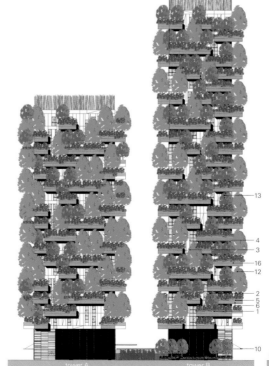

西立面 west elevation

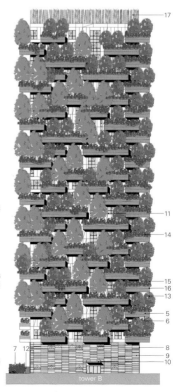

南立面 south elevation

©International High Rise Award (courtesy of the architect)

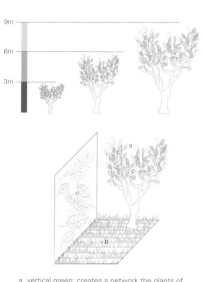

a. vertical green: creates a network the plants of different levels increasing vertical eco functionality
b. horizontal green: diversification and flowering
c. vertical green: forest effect and shading

垂直森林综合体
vertical forest synthesis

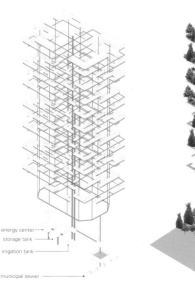

供水系统
water supply system

植被
vegetation

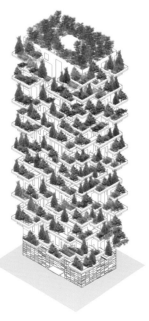

垂直森林
vertical forest

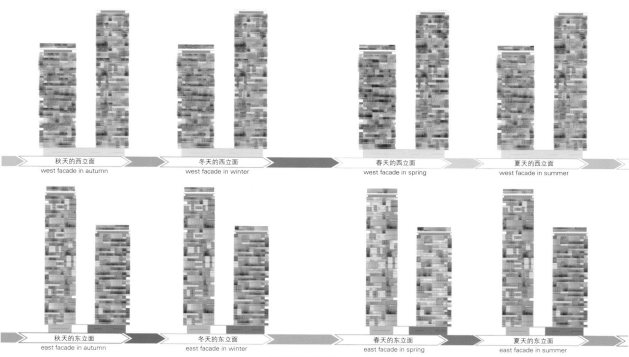

| 秋天的西立面 | 冬天的西立面 | 春天的西立面 | 夏天的西立面 |
| west facade in autumn | west facade in winter | west facade in spring | west facade in summer |

| 秋天的东立面 | 冬天的东立面 | 春天的东立面 | 夏天的东立面 |
| east facade in autumn | east facade in winter | east facade in spring | east facade in summer |

季节性环境示意图
seasonal environmental diagram

ronmental sustainability.

The Vertical Forest increases biodiversity. It promotes the formation of an urban ecosystem where various plant types create a separate vertical environment, able to be inhabited by birds and insects (with an initial estimate of 1,600 specimens of birds and butterflies). In this way, it constitutes a spontaneous factor for repopulating the city's flora and fauna.

The Vertical Forest helps build a micro-climate and filter fine particles contained in the urban environment. The diversity of plants helps develop the micro-climate which produces humidity, absorbs CO_2 and particles, produces oxygen, and protects against radiation and noise pollution.

The Vertical Forest is an anti-sprawl method which helps control and reduce urban expansion. In terms of urban density, each tower constitutes the equivalent of a area of single family houses and buildings around 50,000m².

The choice of species and their distribution according to the orientation and height of facades are the result of three years studies car-

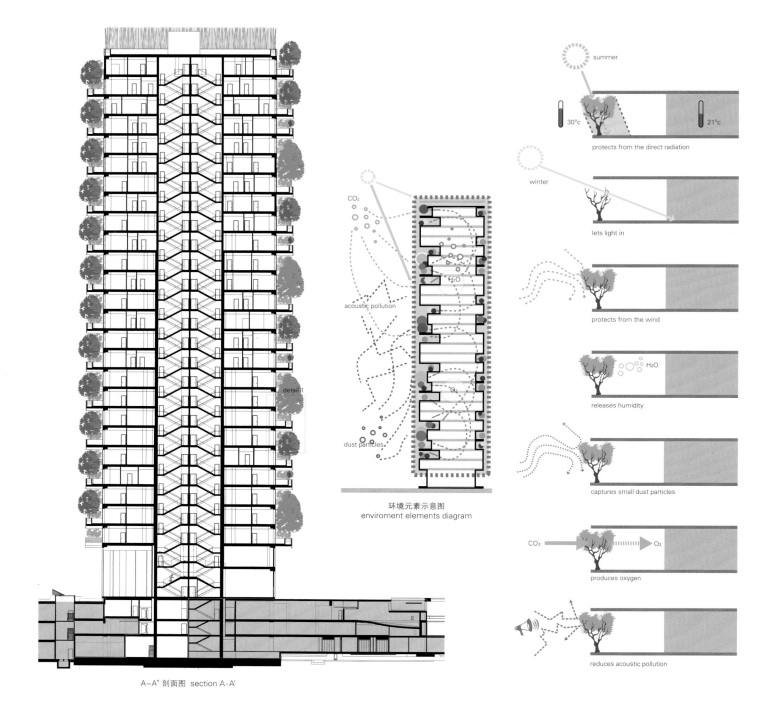

环境元素示意图 enviroment elements diagram

A-A' 剖面图 section A-A'

ried out by a group of botanists and ethologists. The plants which are used on the building were precultivated in a nursery in order for them to become accustomed to similar conditions to those which they will find on the balconies.

The Vertical Forest is an ever-evolving landmark of the city, of which colors change depending on the season and different natures of the plants used. This offers Milan's population an ever-changing view of the city.

The management of the basins where the plants grow is the responsibility of the condominium, as is the maintenance and replacement of all vegetations and the number of plants established for each basin.

Following micro-meteorological studies, the calculation of irrigation requirements was carried out by climatic characteristics and was diversified depending on the exposure of each facade and the distribution of vegetation on each floor.

详图a detail a

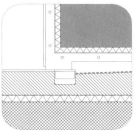

详图b detail b

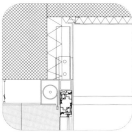

详图c detail c

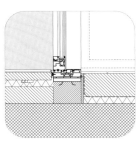

详图d detail d

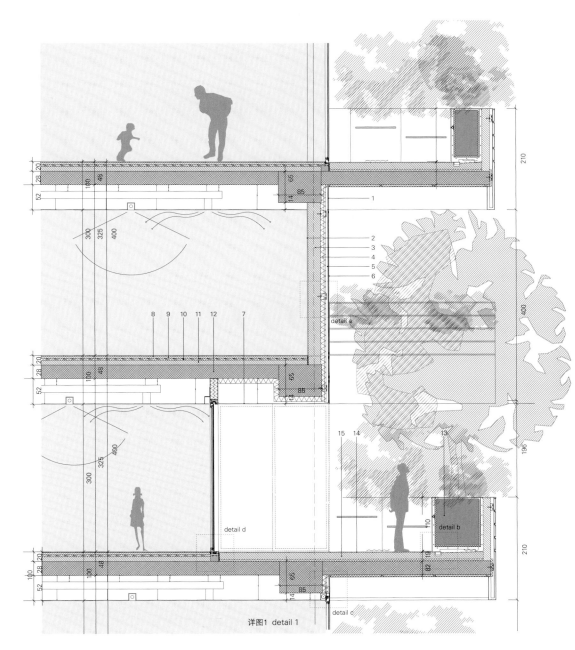

详图1 detail 1

1. facade finish for the covered part of the terraces in composite material composed of the slabs of aluminum in gray anthracite color with translucent effect 2. internal coating 3. masonry in alveolar bricks 4. heat insulator 5. metal substructure supporting ventilated facade 6. ventilated facade with finish in stoneware slabs of gray anthracite translucent color 7. suspended ceiling in plasterboard to hide the beams where the corner of the building is blunt, painted in the color of veil 8. layer of the coating of the floor 9. finishing screed 10. radiant panels 11. engineering plants screed 12. post tense structural ceiling 13. tank prefabricated and assembled on building site composed of the union of the elements with U section 14. finishing in stoneware of gray color RAL 7040 15. lightened screed

项目名称：Vertical Forest / 地点：Milan, Italy
建筑师：Stefano Boeri, Gianandrea Barreca, Giovanni La Varra
垂直森林景观设计：Emanuela Borio, Laura Gatti
美学设计：Davor Popovic, Francesco de Felice
开发商：Hines Italia
设计开发：Gianni Bertoldi_coordinator, Alessandro Agosti, Andrea Casetto, Matteo Colognese, Angela Parrozzani, Stefano Onnis
规划设计和二期设计：: coordinator_Frederic de Smet, Daniele Barillari, Marco Brega, Julien Boitard, Matilde Cassani, Andrea Casetto, Francesca Cesa Bianchi, Inge Lengwenus, Corrado Longa, Eleanna Kotsikou, Matteo Marzi, Emanuela Messina, Andrea Sellanes
结构工程师：Arup Italia s.r.l. / 设施设计：Deerns Italia s.p.a.
细节设计：Tekne s.p.a. / 景观设计：Land s.r.l.
基础设施设计：Alpina s.p.a / 项目和施工管理：Hines Italia s.r.l.
建筑管理和监督：(DL) 2008~2012: MI.PR.AV. s.r.l.
用地面积：40,000m² / 总建筑面积：1,200m² / 有效楼层面积：19,000m²
设计时间：2006~2008 / 施工时间：2008~2013
摄影师：©Paolo Rosselli (courtesy of the architect) (except as noted)

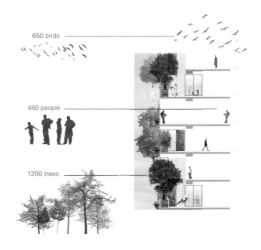

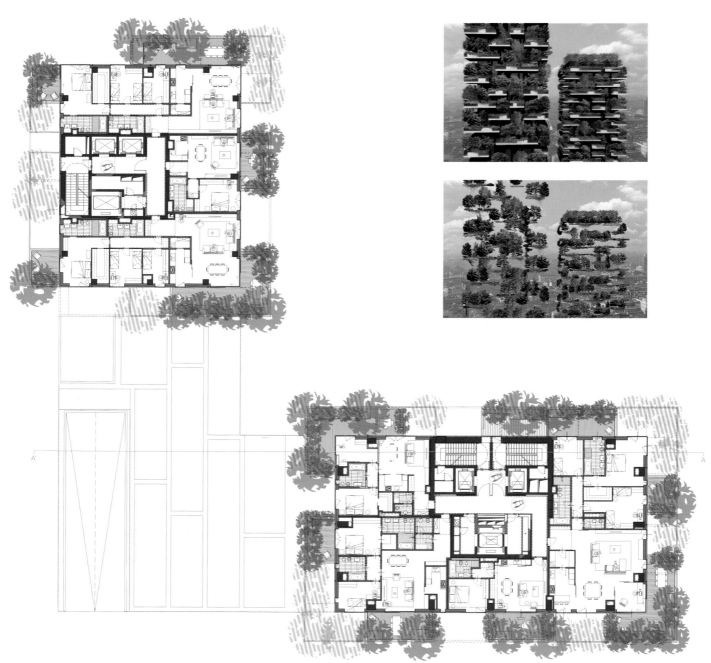

标准层 typical floor

城市住宅
Urban How

Popular and Public

流行化和大众化

人们认为建筑设计是公众的事情，因为建筑物内会有很多政治事件发生，同时它也是人们公开讨论问题的礼堂。但是建筑，作为一处简洁大方的、类似公共场合的对公众开放的空间，首先应该受欢迎，被许多人认可和赞美，另外必须通过外形和功能来提高公共生活质量。因此，对建筑来说，以前从未像现在这样让大众参与到建筑设计中来。这样做有两层含义：一是欢迎观众，即建筑的使用者和居民，同时又不仅仅局限于使用者和居民；二是为公众空间即政治和公共领域提供场所。

建筑面向大众。或者说建筑设计中蕴含的大众理念以及人们对建筑如何塑造公众生活的客观环境引起了大家的关注。首先，人们对公众空间的消失普遍表示关切，不仅因为它们越来越私有化，还因为它们越来越复杂，档次越来越低，慢慢变成了垃圾空间而无人问津了。其次，正如任何事物都有两面性，据说公共机构的建筑变得越来越壮观，但不确定对公共空间质量来说是得还是失。

在建筑设计评论家迪耶·萨迪奇最近出版的一本书中写道：比如说，博物馆是最引人注目的建筑之一，然而它不再是保护和展示珍贵文物和国家宝藏的地方，而变成了人们休闲之所，成为城市改造之所，成为博物馆展示自身的壮观华丽之所。建筑物壮观华丽这件事本身并不是坏事，但是随着以盈利为目的展示，博物馆以及其他公共机构开始成为国际品牌，建筑物成了它们的品牌。世界各地的大城市、小城镇都知道品牌式建筑造成的名牌效应有多大，都希望成功地加盟其中，在城市的中心矗立一栋这样的建筑。我们不知道那些城市改造是否在深层次上正改变着城

Architecture has always been a public affair, even before the notion of the public, both as the realm in which the political takes place or as an auditorium where common matters of public discussion happen. But one thing is to stand there in a space that is open for everyone which is the simplest and less precise definition of public space, to be a building that may be popular, recognized and admired by a lot of people, and another is to enhance the public life by way both of its shape and its performance. Therefore, to engage the public has for architecture, nowadays probably more than ever, a double sense: to invite the audience, a term that for a building would imply its users and inhabitants but would not be reduced to them, to open a space where the public, that is, the political and the civic realms, can take place.

Architecture has gone to public. Or rather, the public implications of architecture and the interest in the way it shapes the physical settings of public life have become in a way matter of public discussion. First it was the commonly expressed concern about the disappearance of public space: not only its obvious privatization but its more complication and degradation: it was becoming junk space. Then, and probably just as the other side of the same coin, public institution architecture was said to be more and more spectacular, without being sure if there was a gain or a loss in the quality of public space.

In a recent book, the design and architectural critic Deyan Sudjic wrote that museums, for instance, one of the most conspicuous one of all our institutions, no longer places served to protect and exhibit valuable objects or national treasures, but had become places for leisure, urban renovation and, overall, spectacular demonstrations of themselves. That a building could be spectacular was not a bad thing by itself. But with the show coming the business, museums, as well as other public institutions, begin to work

阿尔比大剧院_Albi Grand Theater/Dominique Perrault Architecture
什切青新爱乐音乐厅_Philharmonic Hall in Szczecin/Barozzi/Veiga
城市文化欧洲区域中心_Euroregional Center for Urban Culture/Atelier d'architecture King Kong
圣马洛文化中心_Saint-Malo Cultural Hub/AS. Architecture Studio
瓦莱塔城门_Valletta City Gate/Renzo Piano Building Workshop
塔德乌什·坎特CRICOTEKA博物馆_CRICOTEKA Museum of Tadeusz Kantor/Wizja sp. z o.o.＋nsMoonStudio
格旦斯克莎士比亚剧场_Gdansk Shakespearean Theater/Rizzi-Pro.Tec.O

流行化和大众化_Popular and Public/Alejandro Hernández Galvez

市的公共空间和市民与公共空间互动的方式。流行的趋势和大众眼光的关系还是没那么容易理解的，就像流行音乐或是电影大片所给人们的启示：最引人入胜、销量最好的文化产品并不一定就能给公众带来更加丰富的体验。最近，巴塞罗那当代艺术博物馆的独立研究项目的前任负责人保罗·B·普雷西亚多对一些博物馆为了盈利而采取的市场营销和增加财政收入策略的重要性表示质疑，质疑这样的做法是否威胁到博物馆作为"重塑公众民主意识的实验室"这一初衷。同样，如同音乐和电影，大众建筑物和文化展览并不一定要保证更深层次的公共性。

当然，博物馆不只是一座建筑，同时它也是一个收藏、保护、研究和展览藏品和举行一系列文化活动的地方。图书馆也是这样，不仅仅是一座建筑，收集了许多藏书，作为一个公共机构，图书馆也是所有举行与书这一文化产品有关的活动的场所。剧院或学校也是这样的建筑。这些机构都需要建筑物来承载活动，需要一个合适的表演舞台。但是这里所说的建筑物的意义并不像看起来那么明显。尤纳·弗里德曼对在建筑物内展览物品的必要性表示质疑，认为有些建筑很宏伟壮观，不容易改变其物理结构。确实，我们可以把一些物品拿到公共广场进行展览。确实不是所有东西，也许不是任何东西都可以拿到广场进行展览，但是有些展品确实可以这样做。我们也可以把图书馆想象成雷·布拉德伯雷小说里提到的由故事和读者构成的社区。戏剧导演、理论家彼得·布鲁克认为只要有演员和观众，任何空的场地都可以看作是舞台。这就是我们所说的公共的含义。观众，从某种意义上来说，是个体的集合体，变得大众化。

上面所提到的关于公共空间的改变，如私有化、档次降低、变得宏伟壮观，这些在本世纪头十年里由于金融危机的影响变得越明显。宏伟壮观被看作是投机的掩饰，引人注目而有悖常理，是公共空间私有化和档次降低的衍生品，是投机的想法，显然与发明和试验毫无关系。

总之，人们的兴趣似乎变化无常。且不说内容或意图——这需要做深度分析——至少在库哈斯推出的"本源"设计理念之前的威尼斯建筑双年展的主题都是不断变化的。人们从对道德伦理的需求多过审美，到

as international brands for which buildings became part of their trade mark image. Small town and big city majors all around the world knew about that the famous effect that a trade mark building could trigger, and looked forward to get a successful franchise and an existing building in the heart of their cities. It was not always sure if those urban transformations have changed in depth the public space of their cities and the way its citizens interact in it. The relationship between the popular and the public sphere was not simple to understand, as pop music or blockbuster films may show: not always the most spectacular and consumed cultural products are the ones that help to shape a richer experience of the public realm. Recently, Paul B. Preciado, former director of the Independent Studies Program at the Contemporary Art Museum in Barcelona, questioned the importance of marketing and financial expansion strategies undertaken by some museums to become more profitable, and if they were a threat to the idea of museums as "laboratories to reinvent the public democratic sphere" or not. Again, as with music or films, popular buildings and cultural exhibitions do not necessarily guarantee a stronger public sphere.

Of course, we know that a museum is not only or first of all a building. It is an institution that at the same time shelters, takes care, studies and exhibits a collection of objects and a series of cultural programs. The same can be said of a library: it is not only a building and a collection of books but all what happens around the book as a cultural object as a public institution. The same for a theater or a school. All of those institutions obviously need buildings to frame the activities they perform, a proper stage for their performance. But, what we mean by buildings may not be as obvious as it seems. Yona Friedman has questioned the necessity of exhibiting objects inside of a building, meaning by that a somewhat monumental, not quite easy to transform physical structure. And it's true: we can take some objects out into the public square. Not everything, maybe; not anything, for sure, but we can. We can also imagine the library as a community of stories and readers, almost like in Ray Bradbury's fiction. And theater director and theorist Peter Brook proposed that any empty space can be seen as a stage, as long as

认为建筑是人们聚集之所、社会和城市不是建筑的结果而是建筑的条件、建筑学的最终目标是建立一个共同点，这之间的发展变化是具有连贯性的。在最近发行的久负盛名的普利茨克奖刊物中，坂茂被人们所认可部分是因为其强调建筑的社会价值的重要性。弗雷·奥托被认可是因为"他的协作精神以及对资源的合理利用的关注"。最近，伊东丰雄、史蒂芬·霍尔和赖尼尔德·葛拉夫——其观点特别犀利——都撰文说，建筑与经济和不平等性之间的关系以及公共建筑应对大众有益这样并未老套的观念已经发生了翻天覆地的变化，现在人们把建筑作为一种金融工具。因此，建筑不能成为，重要的是不想成为，"重塑公众民主意识的实验室"，正如普雷希亚多提到博物馆应该是重建公众民主意识的地方。最初，这些言论都可以看成与经济相关，但现在显而易见是政治问题，因此也成为公众问题。

本期所呈现给读者的建筑项目有相似特点，也有具体情况。Dominique Perrault建筑事务所设计的阿尔比大剧院（法国）集电影院、剧院、餐厅、商店及其他设施于一体。剧院的大厅位于封闭的礼堂和大剧院简洁的几何玻璃幕墙之间，就像是大剧院外面的露天广场的延伸体，成为室内的公共空间。玻璃幕墙外面的金属网体现了这栋建筑的特色，具有纪念碑式的意义，同时又不失建筑的正式感与朴素感。最近，由Barozzi/Veiga设计的波兰什切青新爱乐音乐厅摘取了2015年密斯·凡·德·罗奖。新爱乐音乐厅有可以容纳1000名观众的交响乐厅和可以容纳200人的室内音乐厅。因为其位于城市具有历史意义的街区，因此为了与此相协调，音乐厅的外表不是透明的，也不像阿尔比大剧院那样更具开放性。不管怎么说，其门厅的空间还是成为开放的公共空间的延续。

格旦斯克莎士比亚剧场（波兰）的内部上空空间是伊丽莎白大剧院。剧院用机械化的屋顶来与天空相连接，四周的外围空间是剧院的各种后勤保障设施，如建筑师所说，同时也作为公共通道。Wizja sp. z o.o.+ns MoonStudio的塔德乌什·坎特CRICOTEKA博物馆设计通过开

the relation between actors and audience is possible. It's all about the public, we may say. The audience, in a sense, or about the way in which that audience, as a collection of individuals, becomes public.

These conditions of the transformation of public space: privatization, degradation and spectacularity, all made more evident as a result of the economical crisis of the first decade of this century. Spectacularity was seen as the seductive but perverse mask of speculation, which derived in privatization and degradation of the public space – an idea of speculation that had nothing to do, obviously, with invention and experimentation.

In any case, interests seem to be shifting, if not for the content or its intentions – which would require more in deep analysis –, at least for the title of each of the former Venice Architectural Biennales before Koolhaas' Fundamentals, and there seems to be a continuity from the demand of more ethics and less esthetics to the idea that people meets in architecture, that society and cities are not a result but a condition of architecture and that the final goal of the discipline is to build a common ground. As for the more recent editions of the prestigious Pritzker Prize, Shigeru Ban was recognized in part for the importance of a social practice of architecture, and Frei Otto for "his collaborative spirit and concern for the careful use of resources". Recent texts by Toyo Ito, Steven Holl and, particularly sharp, by Reinier de Graaf, argue that the relation of architecture to economy and inequality, and the not so old idea of a public architecture that could help to the common good, have dramatically changed to a way of understanding architecture just as a financial instrument. An architecture, therefore, that can not, but most importantly, doesn't want to serve as "laboratories to reinvent the public democratic sphere", as Preciado put it for Museums. At first all those arguments could be read as economical, but there are clearly political, and therefore public issues at sake.

The projects presented in this issue have some similar features and ways of working within their specific contexts. Dominique Perrault Architecture's Albi Grand Theater, combines cinemas and a theater with a restaurant, shops and other facilities. Between the enclosed space of the auditorium and the simple, geometric glass facade, the foyer acts as an internal public space almost like a continuation of the plaza outside the Theater. A metallic screen outside the glass facade conveys the monumental character of the building without compromising the formal austerity of the box.

Barozzi / Veiga's Philharmonic Hall in Szczecin, Poland, recently awarded with the Mies van der Rohe Prize, houses a symphony hall for 1,000 spectators and a music chamber for 200. Probably re-

放的广场上面的反射天花板使人们思考演员和观众的关系。由Atelier d'architecture King Kong设计的位于法国里尔市的城市文化欧洲区域中心名副其实,其开展的文化活动延伸到街上的公共空间。在完成第一阶段的修复改造后,新增加的结构覆盖了透明的玻璃幕墙,另外用锈迹斑斑的金属网将其与街道隔离开来,而金属网与玻璃建筑之间的空间又可成为建筑的室外空间。

位于马耳他的瓦莱塔城门由伦佐·皮亚诺建筑工作室设计。该设计重新定义了作为城市壁垒的石墙的沉重感,在歌剧院的遗址上修建公共建筑,重构这个"城市之门"的城市文脉。由AS建筑工作室设计的圣马洛文化中心位于圣马洛市的市中心,在陆地与海之间,两条相互缠绕的弧形条带将新高铁站同城市历史轴线连接起来,建造了一个公共文化场所并向公共空间开放。

上述这些建筑都是介体,这样说不仅仅因为它们都是举行不同文化活动的舞台,使文化活动吸引更多的观众,甚至说构建新的公共空间,还因为无论从实际情况还是从建筑角度来说,这些建筑物都不仅仅只是结构框架,从某些方面来说还是建筑物内举办的活动与可能在建筑物外举办的活动之间的界面。建筑物周边的街道和广场和建筑物本身的功能一样,都是重要的建筑设计元素。很多建筑师经常使用"象征性的"这个词来描述自己的项目设计,但是他们不认为有纪念意义等同于壮观华丽。无论是设计图纸还是实景,他们的设计既具有象征性又恬静,向城市开放,同时重构城市文脉。这里所说的公众参与有双重含义:一是邀请观众,二是设想一个无论是从建筑自身来说还是从我们来说都不容忽视的一种情况。

sponding to its situation in the corner of a block near to the historic part of the city, it is more opaque and seems less open than Albi Grand Theater. Anyway, it is again the space of the foyer that works as a continuation of the open public space.
In Gdansk Shakespearean Theater, the inner void of the Elizabethan Theater, that can be open to the sky by a mechanical roof, is enclosed by the built facilites and services for the theater that work, as the architect states, as public passageways.
The CIRCOTEKA Museum of Tadeusz Kantor, by Wizja sp. z o.o. + nsMoonStudio, reflects on the relation between actor and audience and it does itself literally with a reflecting ceiling over an open plaza.
The Euroregioinal Center of Urban Cultures in Lille, by Atelier d'architecture King Kong, does exactly what its name suggests, opening cultural practices to the public space of the street. Completing a first phase rehabilitation of the site, the new volumes are cladded in transparent glass and rusted metal mesh is set back from the street to open up a space to be used also outside of the building.
Valletta City Gate, in Malta, by Renzo Piano Building Workshop, reworks the heaviness of the stone walls of the city ramparts and occupies the ruins of a former Opera House with public buildings, reconfiguring the urban context of the City Gate. And the Saint-Malo Cultural Hub by AS. Architecture Studio, a cultural hub in the center of Saint-Malo, in a site between land and sea, connects the new TGV terminal to the historical axis of the city with two intertwined curved strips housing the public program and opening to the public space.
All of these buildings work as mediators, not only because they are the stages for different cultural activities that will look to broaden its audience or even build a new public for their own, but because physically, architecturally, they act not only as frames but in a way as interfaces between the activities that take place inside and the potential activities they can foster outside, and the space of the street and the plaza become as important elements of its architecture as that assigned to the program. In many of the descriptions that the architects make of this projects, the word symbolic is used, but they do not play the game of monumentality as spectacularity. Rather, they make of their own performance a symbolic yet serene gesture that works not only at the iconographic but at street level, opening to and reshaping the urban context. Engaging the public has here a double sense: inviting the audience and assuming a condition that architecture, for its own sake and ours, may no longer neglect. Alejandro Hernández Galvez

阿尔比大剧院

Dominique Perrault Architecture

阿尔比大剧院设计获得2009年国际建筑设计大赛冠军。该项目旨在在表演艺术领域加强其原产地和当代创作的传播,使其现代化。该剧院位于市中心,占地面积很大,包括一座900个座位的礼堂、250个座位的排练大厅、行政和后勤区、各个入口大厅和一个独立的餐厅;还包括八个放映室的电影院、高清投影室和一个380个车位的独立地下停车场。整个地区的城市规划还包括剧院广场、阿塔诺广场和拉彼鲁兹花园。在考虑建筑设计相关问题之前,项目的根本,即初衷是使这座规模重大的大剧院不会改变其城市的实质,来展现阿尔比市市中心一个真实的文化区。

考虑到项目的连贯性和位置,设计师建造了一个"城市走廊",沿途不同的文化建筑和公共空间鳞次栉比,点缀着从大教堂到罗什古德公园的通道。阿尔比市的历史身份最近得到联合国教科文组织的推动,大剧院坐落在这一城市历史中心的边缘,是这座城市的现代性与历史性的完美融合。大剧院位于整个地块的后部,释放了前部的大型空间来作为公共空间,成为宝贵的城市空间,仍具有市中心的地位。

大剧院简约而纯粹的几何外观垂直分布着不同功能,并留出一层平面空间,为公众提供一处宽敞而连续的公共空间。

阿尔比剧院的外立面呈暖色调,从褐色到橙红色各不相同,让人们联想到阿尔比当地生产的砖块。670块铜彩铝饰板装饰着剧院,而一面金属制成的面纱包裹着建筑垂直的玻璃面,产生摩尔纹,使整栋建筑看起来不是那么厚重。静止的网格形成的曲线及反曲线使建筑充满自由、愉悦的象征和诗意。我们也可把它看作是挥洒而下的空中幕布,歌剧院的舞台呼之欲出。犹如蕾丝一般的金属编织的外表皮保护了剧院的功能,同

时又不会将城市的功能隔离,它过滤阳光,且抵挡风雨。这个巨大的装饰物还具有可持续的特色,调节其覆盖的空间及用途。

此项目通过在古铜色金属网下面"渗透"或延展可能的公共空间来将剧院的纵向动态与城市整体的水平运动结合起来。我们通过客观对待营造公共空间与构建文化设施之间的对立辩证关系来消除建筑内外之间的界限。这种态度会让我们收获更多。

Albi Grand Theater

Awarded winner of the international architecture competition for the Albi Grand Theater in 2009, the project aims to modernize and strengthen the place of production and dissemination of contemporary creation in the field of performing arts. On a large site area at the heart of the city, the grand theater comprises a 900 seats auditorium, an experimental hall of 250 seats, administration and logistic area, foyers and an independent restaurant. It also includes a cinematographic complex with eight movie theaters, high definition projection rooms, and an independent underground parking for 380 cars. The urban planning of the area covering the Theater square, Athanor square and Laperouse garden is part of the program as well. At the root of our project, and before any architectural consideration, there is the fundamental desire that the implementation of

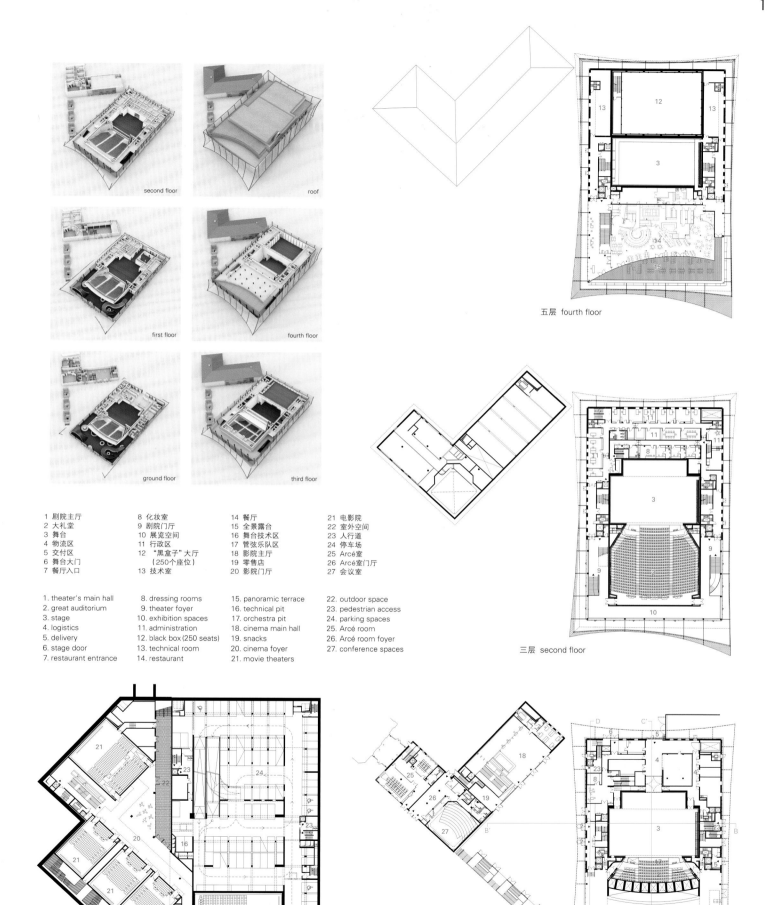

项目名称：Albi Grand Theater
地点：Place de l'Amitié entre les Peuples, Albi, France
建筑师：Dominique Perrault Architecture
当地建筑师：Christian Astruc Architects
结构工程师：VP GREEN / 机械工程师：ETCO
经济学家：RPO / 透视图设计：Changement à vue
音效设计：Jean-Paul Lamoureux / 甲方：Albi Town Council
功能：grand theater, cinematographic complex, independent underground parking, urban planing
用地面积：34,000m²
有效楼层面积：grand theater _ 10,200m², cinema _ 7,800m², independent underground parking _ 12,800m²
设计时间：2009.6 / 施工时间：2011.6—2014.2
摄影师：
©Vincent Boutin (courtesy of the architect) - p.94~95, p.96~97, p.102
©Georges Fessy (courtesy of the architect) - p.98, p.100, p.101, p.103, p.107

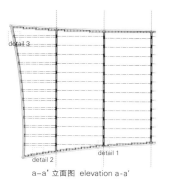

1. golden anodized mesh
2. stainless steel cable-straps
3. stainless steel flat bar
4. flat slot
5. rod
6. socket head screw
7. cable border
8. clevis with external threads
9. clevis with double external
10. threads

a-a' 立面图 elevation a-a'

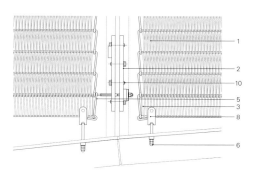

详图1 detail 1

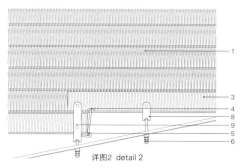

详图2 detail 2

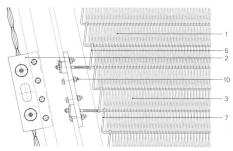

详图3 detail 3

the Grand theater with its significant dimensions does not hurt the urban substance of the city and reveals the existence of a genuine area of culture in the city center of Albi.

The compactness and position of the project on the plot organize and materialize an "urban walk" led by the succession of several cultural buildings and public spaces, arranged in series from the cathedral to Rochegude park. On the outskirts of the historic center, the grand theater identifies a contemporary and complementary counterpoint to the city whose history was recently promoted by UNESCO. Seated at the rear of the plot, the theater liberates large areas for public space at the front, forming a precious urban void which allows the presence of such a piece without strangling the city center's urbanity.

The very simple and pure geometry of the theater organizes vertically the different functions in order to free up the plan of the ground floor to offer a generous and continuous public space.

The warm colors of the facades vary from tan to red-orange, evoking Albi's local brick. 670 copper-colored aluminum panels adorn

the theater. A metallic veil dresses up the prism of the theater. The metal mesh dramatizes much as it delights the general mass of the building, generating "moiré" effects. The curves and counter curves required for this static mesh create a free, happy and lyric architecture. We also see there the metaphor of the drop cloth or the evocation of an opera stage design. This metallic woven skin, as a lace, protects the functions of the theater without separating them from the functions of the city. It filters the light and breaks the wind and the rain. This big ornament has some sustainable qualities. The mesh adjusts to the spaces and the uses that it covers.

The project connects the vertical dynamics of the theater with the horizontal movements of the whole city by infiltrating and extending under the bronzed metal mesh a democratic ground available. We wanted to erase the line between the outside and the inside, taking distance with the dialectic of separation between the practice of public space and that of the cultural facility, an attitude that allows more than it prevents. Dominique Perrault Architecture

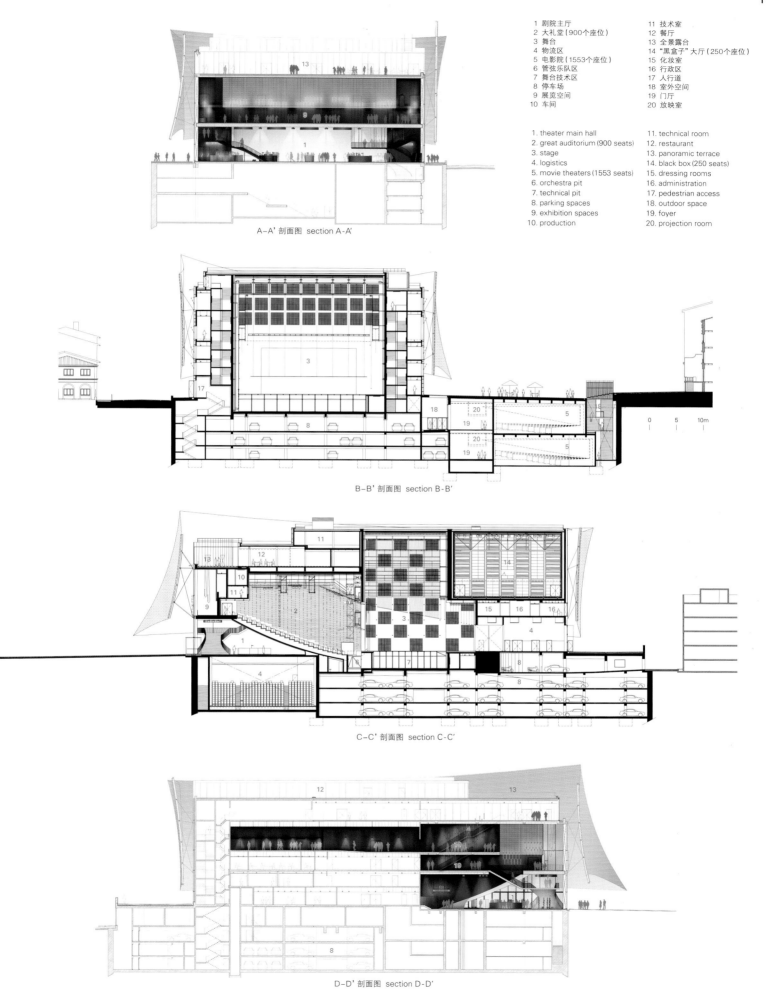

A-A' 剖面图 section A-A'

B-B' 剖面图 section B-B'

C-C' 剖面图 section C-C'

D-D' 剖面图 section D-D'

1 剧院主厅
2 大礼堂（900个座位）
3 舞台
4 物流区
5 电影院（1553个座位）
6 管弦乐队区
7 舞台技术区
8 停车场
9 展览空间
10 车间
11 技术室
12 餐厅
13 全景露台
14 "黑盒子"大厅（250个座位）
15 化妆室
16 行政区
17 人行道
18 室外空间
19 门厅
20 放映室

1. theater main hall
2. great auditorium (900 seats)
3. stage
4. logistics
5. movie theaters (1553 seats)
6. orchestra pit
7. technical pit
8. parking spaces
9. exhibition spaces
10. production
11. technical room
12. restaurant
13. panoramic terrace
14. black box (250 seats)
15. dressing rooms
16. administration
17. pedestrian access
18. outdoor space
19. foyer
20. projection room

什切青新爱乐音乐厅

Barozzi/Veiga

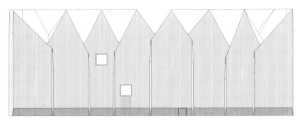

西立面 west elevation

什切青新爱乐音乐厅位于靠近历史老城的城市一角，这里是第二次世界大战中被摧毁的老音乐厅的历史遗址。新音乐厅建筑包括一个可容纳1000名观众席的交响乐大厅、一个可容纳200名观众席的室内音乐大厅、一个用于展览和举行会议的多功能空间，以及可用于举办各种活动的宽敞的门厅。

此建筑看起来是个组合体，同时又是一座综合体，一条连续的公共长廊将各层所有的功能空间串联起来。从外部看，新音乐厅矗立在古老的建筑群中，垂直且呈几何形状的尖屋顶成为主导元素。这些特征使爱乐音乐厅从周围的环境中脱颖而出。

设计的平面构图是由周边的环形边界定义的。这里主要是服务区域。这种做法一方面可以为交响乐大厅和室内音乐大厅提供更大的空间，另一方面定义了建筑与周围环境的关系。一系列尖屋顶呈现出唯一的表现元素，使音乐厅的整体体量得以融入周围城市的细碎轮廓中。

从材料的使用来看，这栋建筑轻盈灵动：玻璃幕墙从内照明，在不同的使用情况下呈现不同的面貌。外立面的朴素和内部流线空间的简洁与主厅内部的表现力对比强烈。按照中欧古典音乐厅的传统，主厅的装修注重装饰物和功能化。该厅的构图依据数学上的斐波那契序列，随着距中心舞台越来越远，天花板上覆盖的黄金叶的碎片不断增加，其形状让人们想起了古典传统。

此建筑主要采用被动式能量控制系统。主要元素是双层外立面，双层外立面的主构件与整体隔音系统和避免高温的自然通风系统的大部分相连。整栋建筑采用LED系统照明，保证建筑物熠熠生辉的同时还能使能源消耗降到最低。与其他区域的屋顶处理不同，为了保证音乐厅隔音和保温系统的最优化，其屋顶也是采用多层次覆盖。

Philharmonic Hall in Szczecin

The new Philharmonic Hall in Szczecin is located on the historical site of the Konzerthaus, which was destroyed during Second World War and recomposes an urban corner in a neighborhood near to the historic city. The building houses a symphony hall for 1,000 spectators, a hall for chamber music for 200 spectators, a multi-functional space for exhibitions and conferences, and a wide foyer, which can also be used to host events.

The building is configured by a synthetic, but at the same time complex volume, which is resolved through a continuous promenade, which connects all these functions through all the levels of the building. Externally, as in the adjacent preexistence, the vertical and geometry of the roof prevail. These characteristics identify the Philharmonic Hall with its surrounding context.

The plan composition is defined by a perimeter ring. This element mostly hosts service spaces. On the one hand this allows to define a large void to gravitate the symphony hall and the hall for chamber music, on the other hand to shape the relationship of

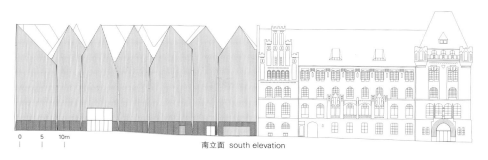

南立面 south elevation

1. plasterboard 2x15mm drywall with galvanised steel system profiles
2. thermal/acoustic 70kg/m³ rockwool insulation
3. profiles galvanised steel sheet. AXL-23 0.75mm or equivalent
4. profiled 0.75mm sheet "minionda 18" or equivalent, online 35mic white finish one sided
5. fluorescent lighting
6. tubular S275 steel truss and tubes. paint finish: 2 coats Hempel Hempadur zinc phosphate primer total coat thickness 40-50 microns(all)+2coats Hempel Hampalux 10 enamel or equivalent white
7. laminated safety low-iron glass 44.2 Vanceva Arctic Snow PVB or equivalent
8. schuco FW 50+AOS or equivalent curtain wall system
9. extruded aluminium louver, polyester>60 microns thermosetting powder coating color RAL white
10. concrete/brick wall
11. Jansen Janisol insulating window system or equivalent, same white finish
12. aluminium 1.2mm sheet, online 35 mic white finish one sided
13. aluminium 1mm sheet, online 35 mic white finish one sided
14. low-e insulating safety glass

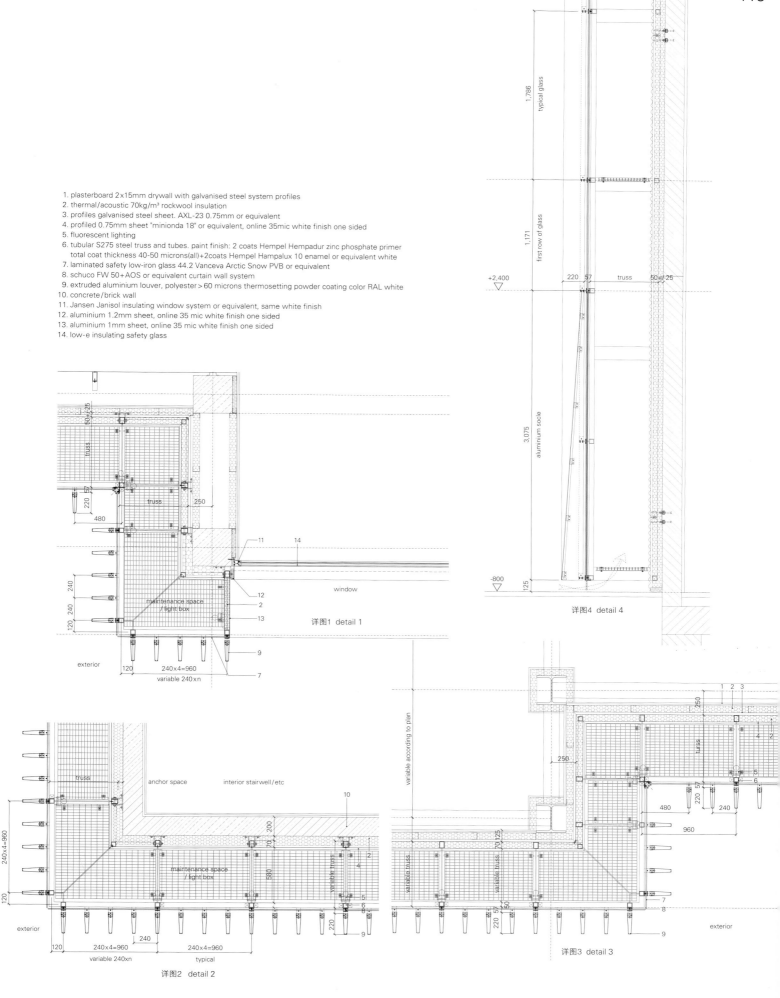

详图1 detail 1
详图2 detail 2
详图3 detail 3
详图4 detail 4

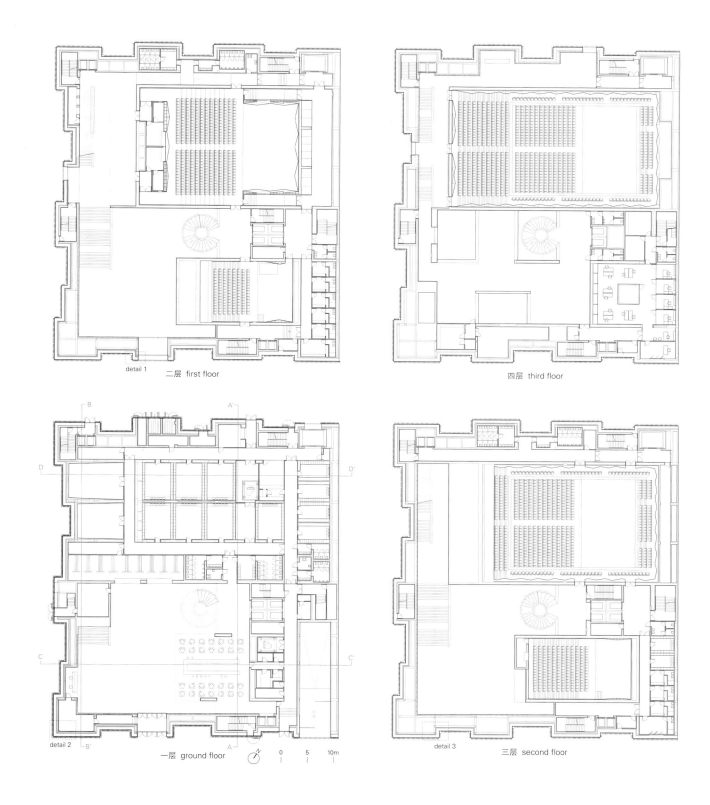

detail 1　二层 first floor

四层 third floor

detail 2　一层 ground floor

detail 3　三层 second floor

the building with its surroundings. The serial modulation of the roof represents the only other expressive element, that permits the integration of the building within the fragmented urban profile of the city.

In its materiality, the building is perceived as a light element: the glass facade, illuminated from inside, depending on the use allows different perceptions. The exterior austerity and the simple composition of the interior circulation spaces contrast with the expressiveness of the main hall. In accordance with the central European tradition of the classical concert halls, decoration becomes ornament and function. The hall is composed following a Fibonacci sequence whose fragmentation increases with the distance from the scene, and gives shape to an ornamental space which reminds of the classical tradition through its gold-leaf covering.

The building predominantly adopts passive systems of energetic control. The main element is the double skin facade channeling a large part of the installation system to provide a global acoustic insulation and a natural ventilation to avoid overheating. Illuminated by a LED system, it turns the building in a glowing volume with a minimum energy consumption. The roof cladding is a multilayer pack, with differences over the concert hall than other zones, to optimize acoustics and thermal insulation.

项目名称：Mieczysław Karłowicz Philharmonic Hall in Szczecin
地点：Małopolska 48, Szczecin. Poland
建筑师：Fabrizio Barozzi, Alberto Veiga
项目负责人：Pieter Janssens, Agnieszka Samsel
项目团队：Marta Grzadziel, Isaac Mayor, Petra Jossen, Cristina Lucena, Cristina Porta, Ruben Sousa
当地建筑师：Studio A4 Sp. z o.o.
结构工程师：Boma S.L., Fort Polska Sp. z o.o.
安装：GLA Engineering Sp. z o.o., Elseco Sp. z o.o., Anoche Iluminación Arquitectónica
音效：Arau Acustica
立面：Ferrés Arquitectos y Consultores
总承包商：Warbud S.A.
用地面积：3,800m²
可用楼层面积：13,000m²
体积：98,200m³
造价：EUR 30,000,000
竞标时间：2007
施工时间：2014
竣工时间：2014
摄影师：©Simon Menges (courtesy of the architect) (except as noted)

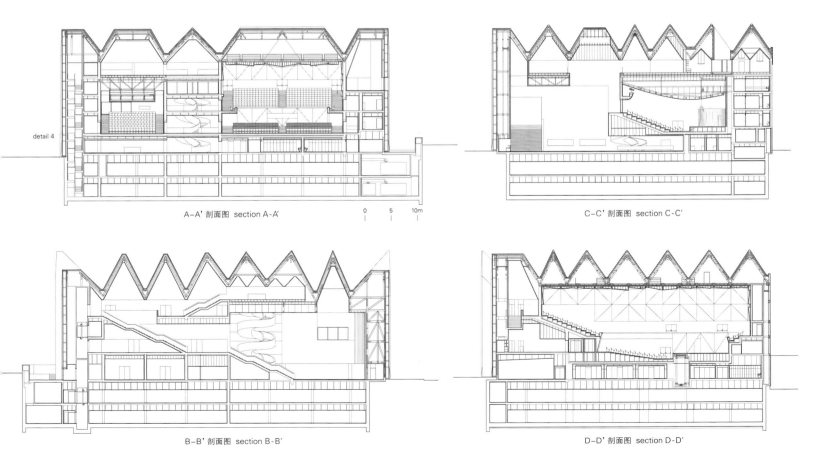

A-A' 剖面图 section A-A'

C-C' 剖面图 section C-C'

B-B' 剖面图 section B-B'

D-D' 剖面图 section D-D'

流动的房屋——欧洲区域城市文化中心——位于里尔市,旨在促进各种各样的街头艺术和表演,为其提供场所。街头艺术形式诞生于城市,该文化中心作为各种不同学科"共同的家",兼具探索城市世界的重任。

设计采用通透的墙壁来唤醒和保持嘻哈风格和街头之间的联系。只有音乐工作室位于地下室,而其余的设施和空间都沿着建筑物的外立面分布。玻璃幕墙(三层玻璃,排气处理)上装有板条百叶窗。嘻哈中心正对着d'Arras街的广场,侧面是位于Dupetit-Thouars路的内部庭院,立面彰显了其作为城市地标的身份。

此建筑共有四层,还有地下室,结构极其简约。每一层对应一个具体的功能区。一楼是接待处和社交娱乐区,可直接通往停车场和前院。地下室用作音乐工作室;二楼是行政办公区;三楼是舞蹈室;四楼是涂鸦工作室,有户外空间可以利用。每一层楼的功能都与使用者的具体需求相得益彰。

城市文化欧洲区域中心
Atelier d'architecture King Kong

项目名称：The Flow/Euroregional Center of Urban Cultures
地点：2 rue de Fontenoy, Lille, France
建筑师：Atelier d'architecture King Kong
合作者：Frdric Neau, Nicolas Broussous
结构/机械/电气工程师：Projex Ingnierie
用地面积：3,400m²
有效楼层面积：4,050m²
设计时间：2010.4
施工时间：2012.4—2014.10
摄影师：
©Julie Soistier (courtesy of the architect) - p.123, p.125
©Roland Halbe - p.118, p.120~121, p.122, p.124, p.127

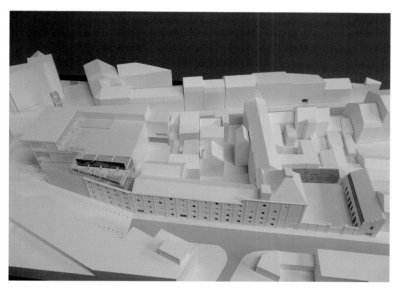

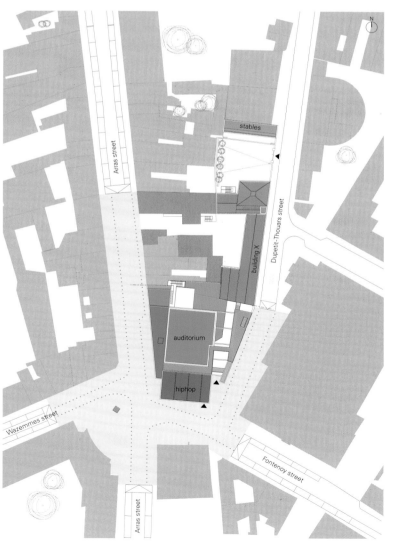

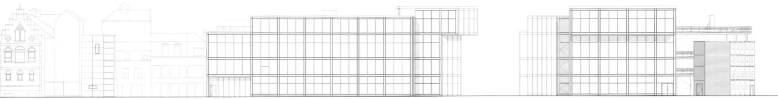

西立面 west elevation　　　　　　　　　南立面 south elevation

东立面 east elevation 0 5 10m

Euroregional Center for Urban Cultures

The Flow – Euroregional Center of Urban Cultures (CECU) in Lille aims both to promote and provide a framework for street art and performance in all its various guises. The center is to function as a home base for a variety of disciplines, while also exploring the urban universe from which such art forms are born.

The design uses transparent walls to evoke and maintain the founding bond between Hip Hop and the street. The basement houses music studios, while the remaining facilities and spaces are distributed along the building's facade. The glazing (triple glazed breathing glazing) is equipped with slatted blinds. The hip hop center faces the square on the Rue d'Arras, flanking the inner courtyard on the Rue Dupetit-Thouars, its facades clearly voicing its presence as an urban landmark.

The building has four floors including the basement and is extremely simply organized. Each floor corresponds to a specific function. The ground floor houses the reception desk and social areas, with direct access to the car park and forecourt. The basement houses the music studios and the first floor the administrative services. The second floor is home to dance and the third floor the graffiti workshop, with outside spaces available. Each floor may therefore function autonomously in harmony with the users' specific requirements.

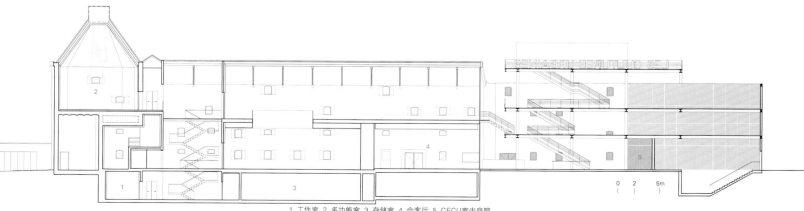

1. 工作室 2. 多功能室 3. 存储室 4. 会客厅 5. CECU室内庭院
1. studio 2. multipurpose room 3. storage 4. meeting hall 5. CECU interior courtyard
A-A' 剖面图 section A-A'

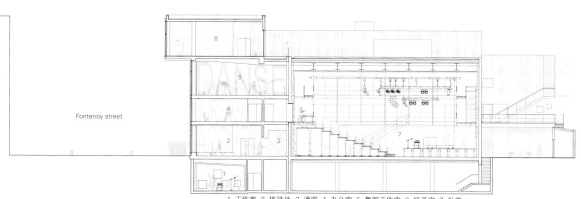

1 工作室 2 接待处 3 酒吧 4 办公室 5 舞蹈工作室 6 绘画室 7 礼堂
1. studio 2. reception 3. bar 4. office 5. dance studio 6. painting space 7. auditorium
B-B' 剖面图 section B-B'

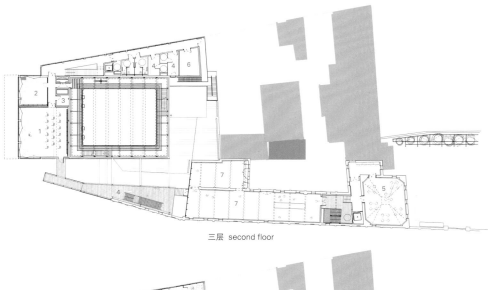

三层 second floor

1 舞蹈工作室
2 集体活动室
3 电气室
4 阳台
5 衣帽间
6 集体活动存储室
7 展览室
8 多功能室

1. dance studio
2. collective activity room
3. electric room
4. balcony
5. locker room
6. collective activity storage
7. exhibition space
8. multipurpose room

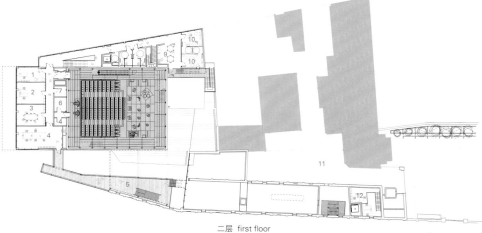

二层 first floor

1 行政办公室
2 通信办公室
3 会议室
4 景观办公室
5 音效室
6 设备存储间
7 机械室
8 阳台
9 备餐间
10 更衣室
11 MFM室内庭院
12 夹层的生产办公室

1. executive office
2. communication office
3. assembly room
4. landscape office
5. sound room
6. equipment storage
7. mechanical room
8. balcony
9. catering
10. dressing room
11. MFM interior courtyard
12. production office mezzanine

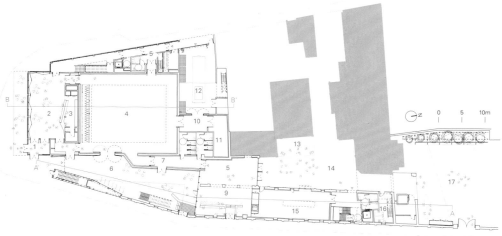

一层 ground floor

1 双层门入口
2 入口大厅
3 酒吧
4 礼堂
5 艺术家通道
6 CECU室内庭院
7 MFM公共入口
8 礼堂
9 会客厅
10 后台
11 存储室
12 服务庭院
13 室外舞台
14 MFM室内庭院
15 展览空间
16 生产办公室
17 绿色庭院

1. double door entrance
2. entrance hall
3. bar
4. auditorium
5. artist entrance
6. CECU interior courtyard
7. MFM public entrance
8. auditorium hall
9. meeting hall
10. back-scene
11. storage
12. service courtyard
13. exterior stage
14. MFM interior courtyard
15. exhibition space
16. production office
17. green courtyard

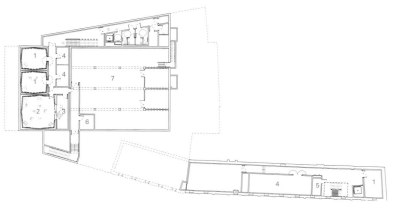

地下一层 first floor below ground

1 工作室
2 录音室
3 控制室
4 存储室
5 清洁间
6 咨询室
7 机械室

1. studio
2. recording studio
3. control room
4. storage
5. cleaning room
6. informatic room
7. mechanical room

圣马洛文化中心

AS. Architecture Studio

圣马洛市的媒体图书馆和艺术电影院的位置具有战略性和象征意义：位于市中心以及将新火车站与城市和大海连为一体的历史轴线上。这一独特的位置使人们走出火车站的时候，第一眼看到的城市地标就是这个文化枢纽。这个项目意在通过这一新的城市中心展现新颖而现代的城市与建筑的呼应。既是建筑又是广场的文化中心旨在成为圣马洛市文化艺术的标志。建筑师通过使用一样的建筑语言和美学，将建筑物的室内、室外空间有机地连为一体。一年之中，不分昼夜，所有的日常活动、文化活动和特殊活动都在这一极具吸引力的场所展开。该项目既融于周围环境又相对独立，既拥有鲜明独特、极具识别性的建筑设计又构成圣马洛市的主要轴线。该项目由两座波浪形建筑物组成，为火车站广场注入一缕动感，是这座城市向大海开放的象征，分隔图书馆和电影厅在两侧的玻璃门厅进一步强调了这一象征。两座文化建筑的屋顶均被绿化植被覆盖，并由一条架空其上的带状光伏太阳能板群相连接。热能由地热系统提供。该项目得到法国NF第三产业建筑高环境质量可持续发展认证，并贴有THPE ENR －30％高节能表现的使用标签。

Saint-Malo Cultural Hub

The location of the media library and art cinema in Saint-Malo is strategic and emblematic: it is situated both in the city centre and on the historical axis that connects the new railway station to the city and the sea. This exceptional location turns the cultural hub into the city's first visible landmark when exiting the railway station. This project implies an original, contemporary urban and architectural response, up to the stakes of the new city centre. As an "esplanade building", it is meant to become Saint-Malo's cultural

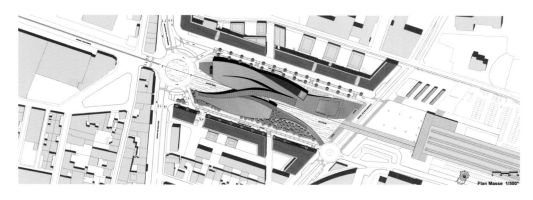

icon. Through the same language and aesthetics, this complex brings together the inside and outside areas. Daily and cultural activities, as well as exceptional events interact in this unique spot, very attractive by day and at night, throughout the year. The project is both contextual and autonomous as it creates an iconic architectural identity and at the same time structures the main axis of Saint-Malo. The building is composed of a double wave that sets the station esplanade in motion. This symbol of a city open onto the sea, is enhanced by a glazed urban foyer that leads to the library on one side, and to the cinema on the other. Both cultural entities are covered with green roofs and bound together by a photovoltaic ribbon. Heat is provided by a geothermal process. This building is certified "NF Bâtiment Tertiaire-démarche HQE" (sustainable design approach for tertiary buildings) and labelled "THPE ENR -30%" (very high energy performance).

东立面 east elevation

1 门厅——"室内街道" 2 媒体图书馆——阅览区 3 环形带状区——员工办公室 4 夹层——影院等候区
5 影院——150个座位 6 控制室 7 技术区 8 安装光伏电池设备的拱形结构
1. foyer – "interior street" 2. media library – reading area 3. looping band – staff offices 4. mezzanine – cinema waiting area
5. 150 seats cinema 6. control room 7. technical areas 8. photovoltaic arch

A-A' 剖面图 section A-A'

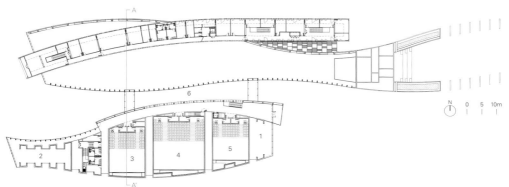

1 充满文化氛围的咖啡厅 2 展览室 3 影院——150个座位 4 影院——220个座位 5 影院——100个座位 6 夹层 7 环形带状区
1. literature cafe 2. exhibition room 3. 150 seats cinema 4. 220 seats cinema
5. 100 seats cinema 6. mezzanine 7. looping band

二层 first floor

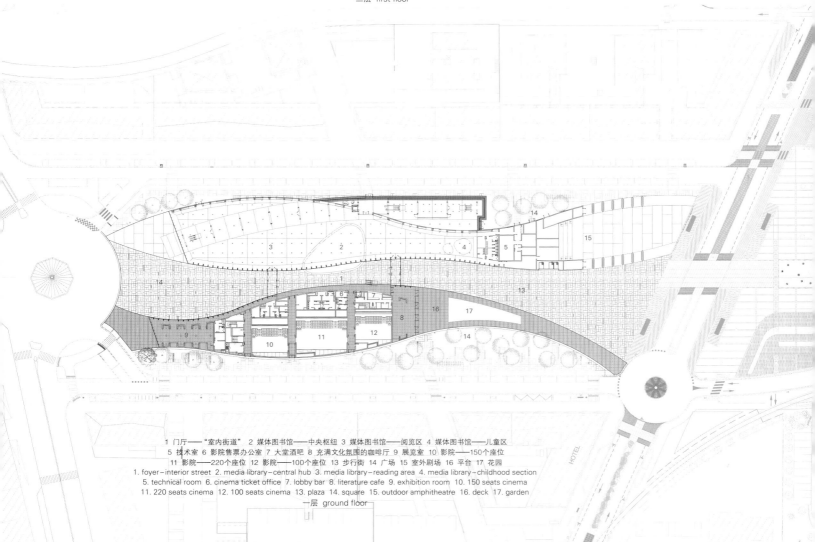

1 门厅——"室内街道" 2 媒体图书馆——中央枢纽 3 媒体图书馆——阅览区 4 媒体图书馆——儿童区
5 技术室 6 影院售票办公室 7 大堂酒吧 8 充满文化氛围的咖啡厅 9 展览室 10 影院——150个座位
11 影院——220个座位 12 影院——100个座位 13 步行街 14 广场 15 室外剧场 16 平台 17 花园
1. foyer – interior street 2. media library – central hub 3. media library – reading area 4. media library – childhood section
5. technical room 6. cinema ticket office 7. lobby bar 8. literature cafe 9. exhibition room 10. 150 seats cinema
11. 220 seats cinema 12. 100 seats cinema 13. plaza 14. square 15. outdoor amphitheatre 16. deck 17. garden

一层 ground floor

项目名称:"La Grande Passerelle" Saint-Malo Cultural Hub / 地点:Saint-Malo, France
设计团队负责人:Architecture-Studio
工程师:Arcoba / 结构/外表皮设计工程师:T/E/S/S / 照明工程师:8'18" / 音效工程师:AVA
功能设计:Le Troisième Pôle / 视觉和标志设计:Thomas Kieffer
特殊家具设计:Agnès Martin, Yves Lamblin _ Omlarchitecture
承包机构:Town of Saint-Malo / 总承包商:SOGEA Bretagne
用地面积:10,224m² / 总建筑面积:6,500m² / 有效楼层面积:4,432m²
竞标时间:2009 / 竣工时间:2014
摄影师:
©Mereglier-Coudrais (courtesy of the architect) - p.128~129
©Luc Boegly (courtesy of the architect) - p.130, p.132, p.133

瓦莱塔城门
Renzo Piano Building Workshop

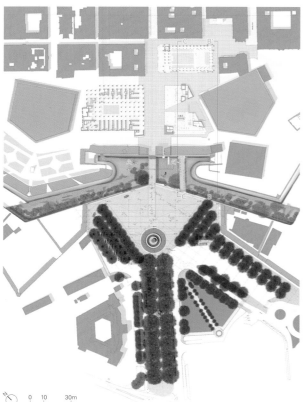

"城门"项目旨在对马耳他首都瓦莱塔的主入口进行全面改造,包括四个部分:城门及其城墙外的改造、前皇家大剧院遗址里的露天剧场"machine"设计、新国会大厦的建造以及护城河的景观美化。

城门、护城河和城墙

项目重点是将桥梁恢复到1633年Dingli城门的规模,对后来的扩建部分进行拆除,使路人能够再次体验过桥的真实体验。瓦莱塔的首个城门,可能是通过城市壁垒的一条单向通道,在历史的岁月里经过多次改建,已经不复其坚固城门的形象。最近一次改建是在50年前,这次改建拆除了32m长的城墙,改变了城市入口的影响力。因此该项目的首要目标是恢复城墙的原始面貌,包括深度和强度,并突出城市入口的狭窄感,同时开放共和国大街的视野。新城门是墙壁上的"缺口",只有8m宽。通过植入60mm厚的钢铁"刀片",新旧墙壁之间有所划分,从而明确区分原来的与已重建的城墙。

城门和护城河之间由楼梯和观光电梯连接,游客可以下到护城河中参观。

新国会大厦,一座环保型建筑

国会大厦由两座巨大而厚重的石头建筑组成,石块与细长的石柱保持了平衡,赋予建筑一种轻盈感,从整体上也尊重了现有街道的布局线。两座石头建筑之间是中心庭院。最北边的那栋主要容纳国会会堂,而南部建筑是国会成员办公室和首相及反对党领袖的办公室的所在地。

国会大厦的立面采用结实的石材作为饰面。这些石材经过精心雕刻,看起来饱经阳光和周围景观朝向的洗礼,其实际功能是过滤太阳辐射,允许自然光线进入室内,同时保证建筑的视野。

建筑物的地下室用来组织某些与国会有关的活动。地下室可通向庭院,庭院内植物繁茂,绿树成荫。这一层也与古老的马耳他铁路隧道相连。这座古老的地下建筑原来被用作车库,现在要恢复原状并向大众开放。

皇家歌剧院遗址

皇家歌剧院由爱德华·米德尔顿·巴里设计,建于1862至1866年间,但于1942年被炸毁。从那以后,废墟就一直作为考古遗迹、场地历史的见证,以及当地集体记忆的碎片。而现在废墟已经被恢复,与露天剧场的新项目整合为一体,且十分安全。在没有举行活动的时候,剧院就是一个开放的露天广场,一处公共的城市空间,也是一个社交场所。夏天,位于历史遗迹内的钢架结构如同一个"剧场加工器"。剧场里可以安装近1000个座位。整个夏季,歌剧、舞蹈、戏剧、音乐会都在这里举行。这需要各种不同的剧院配置,但有了机械化的舞台和剧场的侧翼建筑,这一切都成为可能。剧场还配有ERES音响增强系统,可以重现室内音乐厅的混响和音质——这是为这一地中海历史名胜独创的一项现代技术。

Valletta City Gate

The "City Gate" project takes in the complete reorganization of the principal entrance to Valletta of Maltese capital. The project comprises four parts: the Valletta City Gate and its site immediately outside the city walls, the design for an open-air theater "machine" within the ruins of the former Royal opera house, the construction of a new Parliament building and the landscaping of the ditch.

The gate, the ditch and the city walls

The project focuses on returning the bridge to its 1633 "Dingli's

Gate" dimensions, by demolishing later additions. This allows passers-by to once again have the sensation of crossing a real bridge. Valletta's first city gate, which was probably a single tunnel through the city's ramparts, has been remodelled through the years, considerably altering the image of a fortified city gate. The most recent modification, completed 50 years ago, involved demolishing 32m of the city wall, distorting the impact of the entrance into the city. The first objective of the project was therefore to reinstate the ramparts' original feeling of depth and strength and to reinforce the narrowness of the entrance to the city, while opening up views of Republic Street. The new city gate is a "breach" in the wall of only 8m wide. The relationship between the original fortifications and those that have been reconstructed is made clear by the insertion of powerful 60mm thick steel "blades" that slice through the wall between old and new.

The gate and the ditch will be linked by a stairway and a lift with panoramic views, allowing visitors to descend to the depths of the ditch.

The Parliament, an environmentally responsible building

The parliament building is made up of two massive blocks in stone that are balanced on slender columns to give the building a sense of lightness, the whole respecting the line of the existing street layout. The two blocks are separated by a central courtyard. The northernmost block is principally given over to the parliament chamber, while the south block accommodates members of parliament's offices and the offices of the Prime Minister and Leader of the Opposition.

The parliament's facades are finished in solid stone. This stone has been sculpted as though eroded by the direction of the sun and the views around it, creating a fully functional device that filters solar radiation while allowing natural daylight inside, all the while maintaining views from the building.

Certain organizations and parliament-related activities will be housed on the building's basement level, which opens onto a planted, and shaded courtyard. The old Malta railway tunnel is also connected to this lower level garden space, restoring the old underground structure that had been used as a garage, and making it accessible to the public.

The Royal Opera House site

Built between 1862 and 1866 by E.M. Barry, the Royal Opera House was destroyed during the bombardments of 1942. Ever since, its ruins have been treated like archaeological remains, evidence of the site's history, and fragments of local collective memory. The ruins have now been restored, made secure and integrated into the new project for an open-air theater. When events are not being held here, the theater functions as an open plaza, a public city space, and a social space. During the summer, the steel structure placed within the historic remains works like a "theater machine". Almost 1,000 seats are installed and a theater takes form for a season of opera, dance, theater and concerts, a program that requires a variety of theater configurations that are made possible thanks to the structure's mechanised stage and theater wings. The theater is also equipped with an ERES acoustic enhancement system that can recreate the reverberations and acoustics of an interior concert hall space – an innovative modern technique for this historic Mediterranean site.

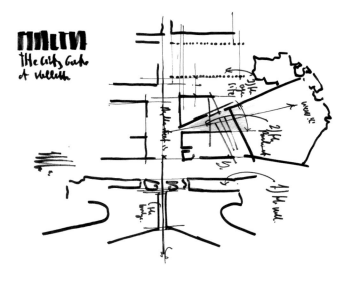

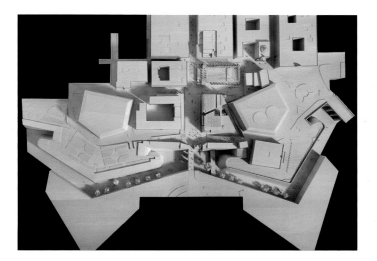

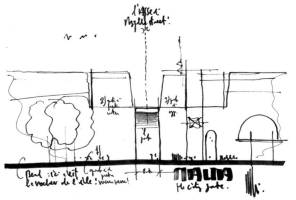

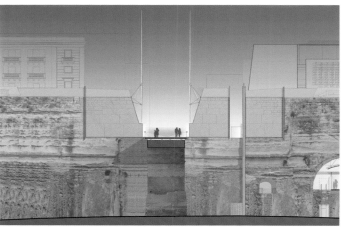

A-A' 剖面图_桥 section A-A'_bridge

B-B'剖面图 section B-B'

1 辩论室 2 国会休息室 3 部长秘书办公室 4 部长办公室 5 歌剧院
1. debating chamber 2. parliament lounge 3. minister secretary's office 4. minister office 5. opera house
C-C' 剖面图 section C-C'

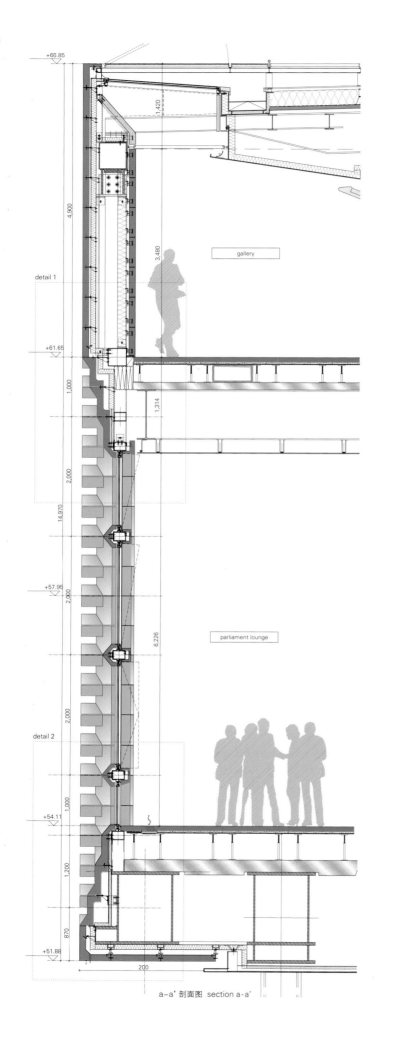

a-a' 剖面图 section a-a'

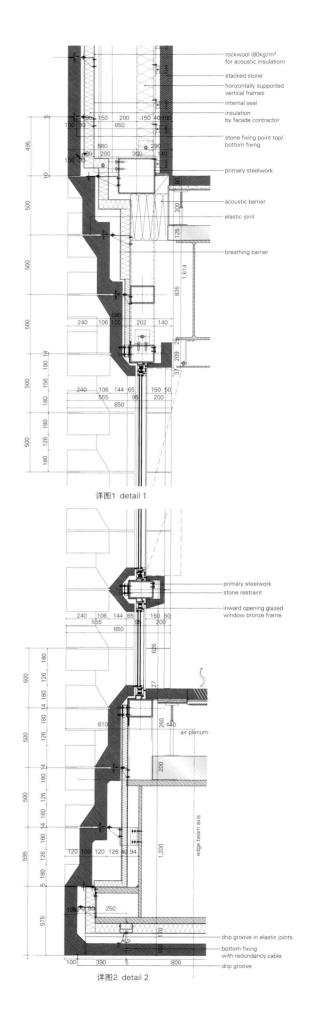

详图1 detail 1

详图2 detail 2

项目名称：Valletta City Gate / 地点：Valletta, Malta
建筑师：Renzo Piano Building Workshop
合作者：Architecture Project (Valletta)
主要合作者：A.Belvedere, B. Plattner
项目团队：D. Franceschin, P. Colonna, P. Pires da Fonte, S. Giorgio-Marrano, N. Baniahmad, A. Boucsein, J. Da Nova, T. Gantner, N. Delevaux, N. Byrelid, R. Tse and B. Alves de Campos, J. LaBoskey, A. Panchasara, A. Thompson, S. Moreau, model _ O. Aubert, C. Colson, Y. Kyrkos
音效/土木/结构/MEP工程师：Arup
石材设计顾问：Kevin Ramsey
剧院顾问：Daniele Abbado
照明：Franck Franjou / 景观：Studio Giorgetta
剧院特殊设备设计：Silvano Cova
甲方：Grand Harbour Regeneration Corporation
用地面积：40,000m²
总建筑面积：parlamento_7,000m², opera_2,800m², venue_1,800m², backstage_1,000m²
高度：19.7m (maximum)
造价：EUR 80,000,000 / 设计时间：2008 / 竣工时间：2015
摄影师：
©Michel Denancé-p.134~135, p.137, p.138, p.140~141, p.143, p.144, p.145, p.148, p.149
©Mario Carrieri (courtesy of the architect)-p.146, p.147

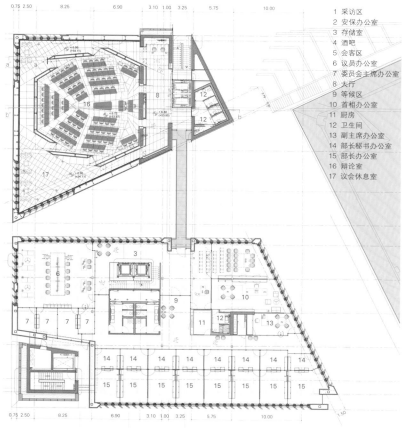

1 采访区
2 安保办公室
3 存储室
4 酒吧
5 会客区
6 议员办公室
7 委员会主席办公室
8 大厅
9 等候区
10 首相办公室
11 厨房
12 卫生间
13 副主席办公室
14 部长秘书办公室
15 部长办公室
16 辩论室
17 议会休息室

二层_议会建筑 first floor_parliament building

1. interview area 2. security office 3. store 4. bar 5. meeting area 6. members of parliament's office
7. committee chairman's office 8. lobby 9. waiting area 10. prime minister's office 11. kitchen 12. toilet
13. deputy chairman's office 14. minister secretary's office 15. minister office 16. debating chamber 17. parliament lounge

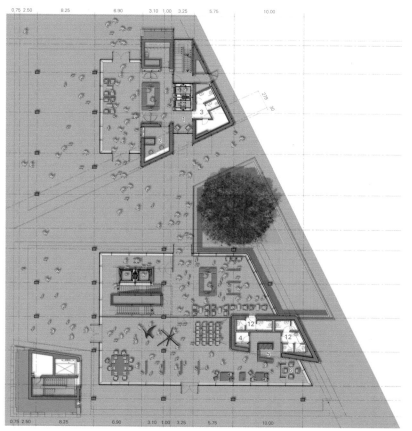

一层_议会建筑 ground floor_parliament building

1. 400mm thk. RC core wall
2. open gutter
3. floor plenum for air distribution
4. 400mm thk. RC wall
5. 300mm thk. RC wall
6. 500mm thk. RC wall
7. Pad foundation for core walls refers to structural drawings
8. 350 x 350mm concrete column
9. concrete beam to support rock
10. old railtrack platform refers to survey drawings acoustic absorbing ceiling w/ plasterboard
11. backing for acoustic insulation
12. stone interior finish air supply in floor plenum
13. below chamber 2 layers hung plasterboard, 50mm mineral wool, 200mm void
14. steel ceiling fixed to primary structrue
15. ground floor facade
16. double glazing with glass fins
17. 650x650 concrete column
18. floor plenum for air distribution steel fabricated
19. column 620 x 400 x 20mm

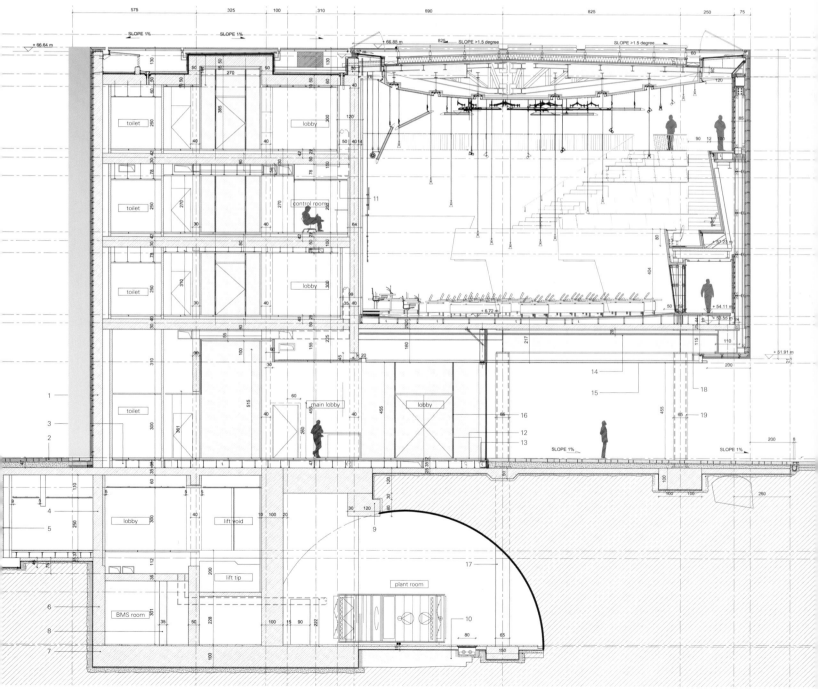

b-b' 剖面图 section b-b'

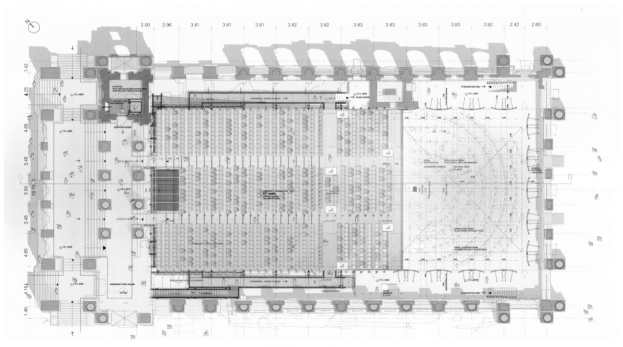

一层_歌剧院 ground floor_opera house

1 入口	7 水管设备间		
2 电梯通道	8 卫生间		
3 走廊	9 聚光平台		
4 舞台指令区	10 照明平台		
5 人行桥	11 水库		
6 超低音音响区	12 舞台		

1. entrance　　7. plant room for plumbing
2. elevator access　8. toilet
3. corridor　　　9. spotlight platform
4. stage direction　10. lighting platform
5. footbridge　　11. reservoir
6. subwoofer　　12. stage

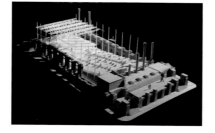 

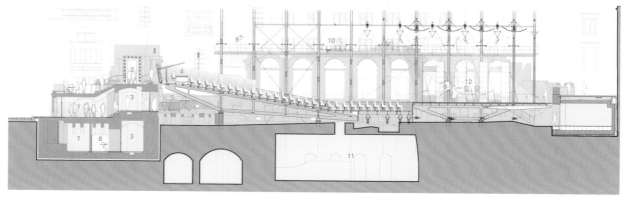

D-D' 剖面图_歌剧院 section D-D'_opera house

城市改造——流行化和大众化 Urban How – Popular and Public

塔德乌什·坎特CRICOTEKA博物馆
Wizja sp. z o.o. + nsMoonStudio

塔德乌什·坎特艺术文档中心"CRICOTEKA"是对个人如何创新进行艺术探索的象征，它跨越了演员和观众之间的界限，跨越了创作者和接受者之间的界限，使每个人都参与到艺术活动-游戏-集体交流这一游戏中来。这一城市公共空间容纳舞台和观众，是每天都举办精彩表演的场所。其价值在于消除了室内外的边界，使空间变得更加流动，群分群享。通过特定的构件、结构、方式和方法等，使创作者、艺术家、居民和游客共同参与空间的营造过程。这也是坎特艺术中心所要宣扬的信息。

该博物馆并不作为一个静态的展示个人艺术作品的场所，而是作为艺术动态视觉的一个延续，打破了传统的条条框框。它的价值不仅在于形式的自主自发性，还在于使艺术家和其工作得以发展的创造性的过程。博物馆所营造出来的空间是为人们行动前做好的"准备"，而这一行动可以随时随地、自然而然地进行。

维斯瓦河、河的两岸和周围环境都是"物品和道具"，都参与到艺术活动-游戏-集体交流这一游戏中来，它们在舞台上的表现（通过一面镜子反射在天花板上）为创造性地使用博物馆提供了条件。

建筑师最初的想法是开发整个地块，在维斯瓦河畔建一个进行演出、展览、表演秀等各种活动的"剧场"，最终放弃了这一想法，使得博物馆的建筑形式和内容的影响范围扩大到整个城市。

后来增加的建筑看上去像一个"神秘"的、被包裹的"物体"，其内部隐藏的是功能性结构。

这一综合体的新建部分需要在材料（钢、钢筋混凝土、玻璃）和静态方面采用非传统的解决方案。建筑物的结构设计源于如下理念：创造一处外部的"剧院"空间，整体布局中的单个元素相互之间产生共同的张力。

CRICOTEKA博物馆由两个交叉的、基于两个支座的钢筋混凝土桁架拱组成。第三个墩柱是双摇臂钢结构。这个分支提供给桁架一个向上的作用，等于荷载和外力下的恒定变形所用的力。由于结构具有复杂的几何系统和其他延伸结构，因此向上的作用力是不同的，这在设计这个交叉结构时便有所考虑。依靠钢筋混凝土的预制支座通过5点承载点和一个晶状体施加作用。上部桁架作用产生的水平力由埋在支座中的特殊锚结构承担。下部的水平力由支座墙体推力依次传递。

作为公共事业机构，"CRICOTEKA"博物馆成为城市的特征之一，或者说是主要景观。建筑物的规模会适当平衡——也可能减少周围的高层建筑的负面影响。博物馆看上去的高度与其背景中建筑物的高度差不多，但远远低于附近新建的酒店的高度。

Cricoteka Museum of Tadeusz Kantor

The Center of Documentation of the Art of Tadeusz Kantor "CRICOTEKA", the symbol of artistic search for the individual path to creativity, crossing the border between the actor and the audience, between the creator and the recipient, engages everyone in an activity-game-collective play. The municipal public space constitutes both the stage and the audience; it is a venue for constant performance. Its value lies in effacing the borders between the inside and the outside, the space "flows", which is shared. The creators, artists, residents and visitors all take part in the process, the activity of creating space which can be shaped with selected objects, structures, means and methods. The Center of Kantor's art should promote this message.

The building is not intended to function as a static presentation of the individual artistic works of the artist but rather as a continuation of his dynamic vision of art breaking away with conventions, valuing it not only for the autonomous character of form but also for the creative process which allows for the development of the artist and his work. The created space is a "preparation" for action, which can take on a happening nature.

The Vistula river, its banks and surroundings are "objects and props" which can take part in the play and their presence on the stage (through a mirror reflection on the ceiling) provides grounds for creative use.

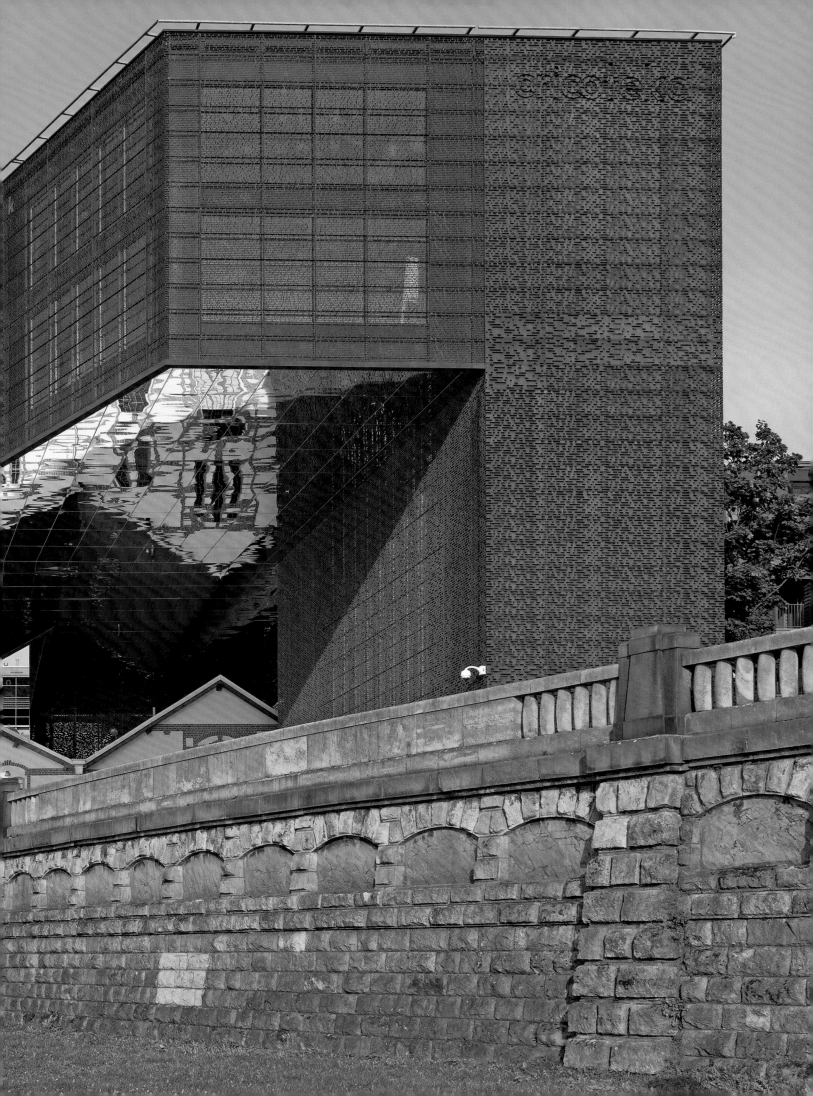

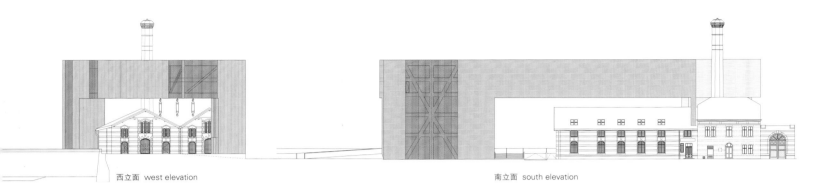

西立面 west elevation   　　　　　　南立面 south elevation

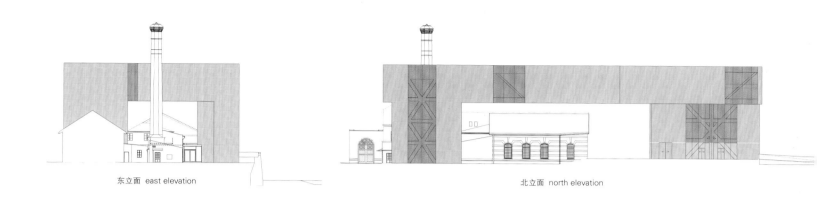

东立面 east elevation   北立面 north elevation

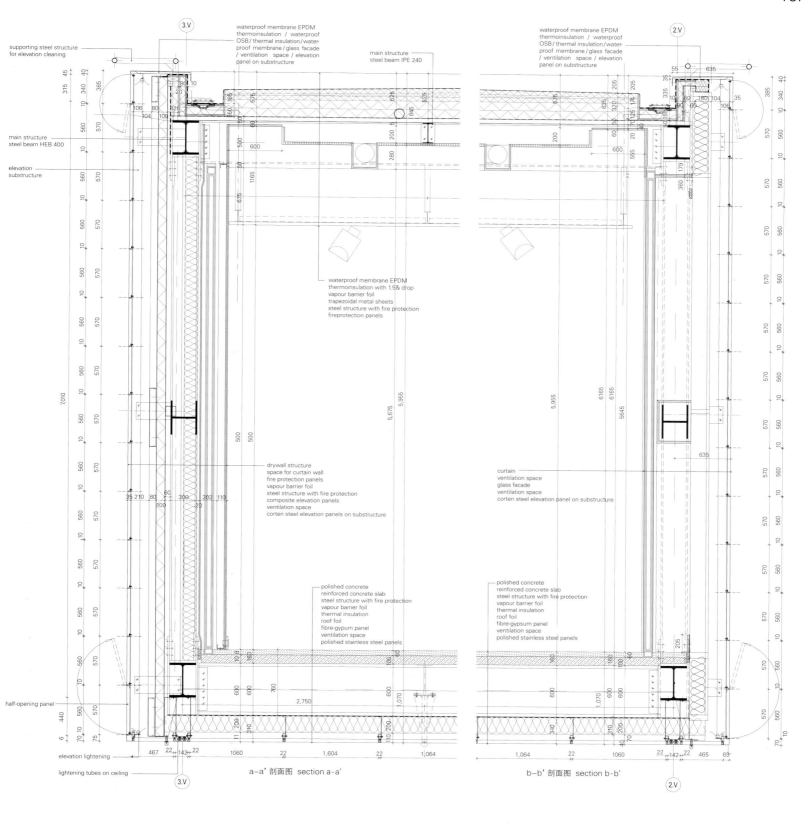

a-a' 剖面图 section a-a'

b-b' 剖面图 section b-b'

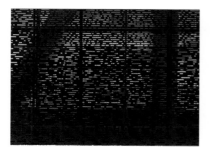

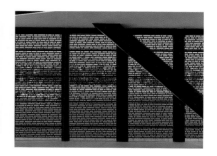

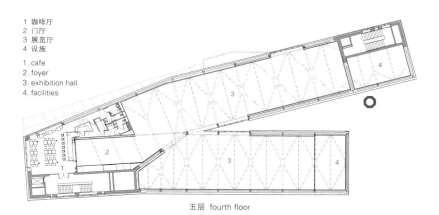

1 咖啡厅
2 门厅
3 展览厅
4 设施

1. cafe
2. foyer
3. exhibition hall
4. facilities

五层 fourth floor

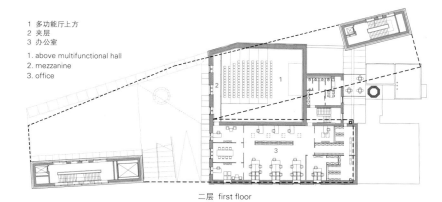

1 多功能厅上方
2 夹层
3 办公室

1. above multifunctional hall
2. mezzanine
3. office

二层 first floor

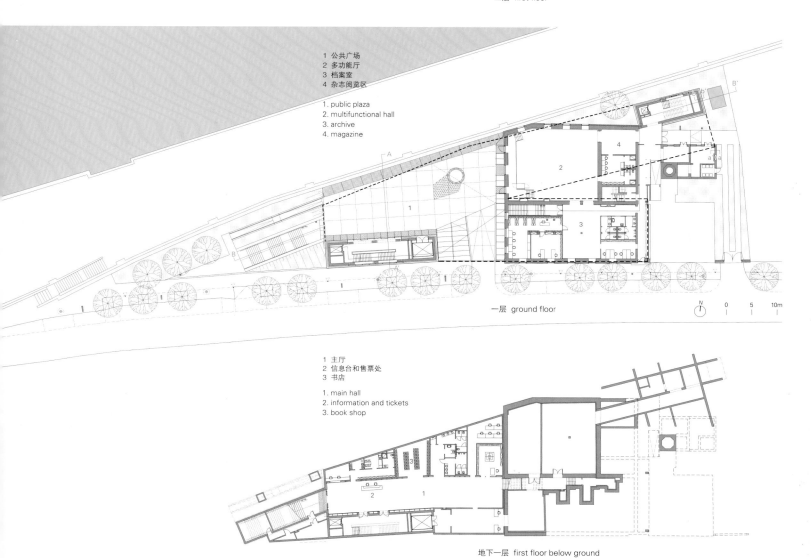

1 公共广场
2 多功能厅
3 档案室
4 杂志阅览区

1. public plaza
2. multifunctional hall
3. archive
4. magazine

一层 ground floor

1 主厅
2 信息台和售票处
3 书店

1. main hall
2. information and tickets
3. book shop

地下一层 first floor below ground

项目名称：CRICOTEKA Museum
地点：Nadwislanska street, Krakow, Poland
建筑师：Stanisław Denko (Wizja sp. z o.o.), Agnieszka Szultk, Piotr Nawara (nsMoonStudio)
合作者：IQ2-Group Consortium
项目团队：Sławomir Zielinski (project leader), Tomasz Gomułka, Michał Marcinkowski, Marcin Kowalewski, Adam Wereszczynski, Marzena Surowiec-Doton, Monika Mackiewicz, Łukasz Skorek, Karol Grec, Katarzyna Ceran, Bartłomiej Łobaziewicz, Ewelina Siestrzewitowska
甲方：The Center of Documentation of the Art of Tadeusz Kantor – CRICOTEKA
用地面积：2,428m² / 总建筑面积：1,646m² / 有效楼层面积：5,265m²
设计时间：2006 / 竣工时间：2014
摄影师：©Wojciech Kryński (courtesy of the architect)

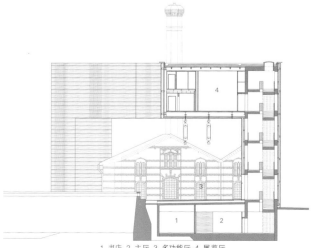

1 书店　2 主厅　3 多功能厅　4 展览厅
1. book shop 2. main hall 3. multifunctional hall 4. exhibition hall
A-A' 剖面图　section A-A'

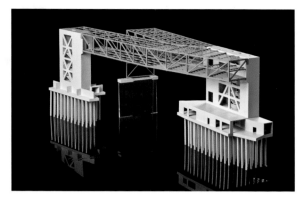
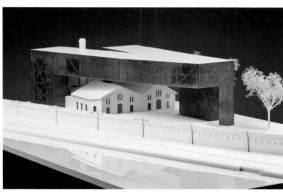

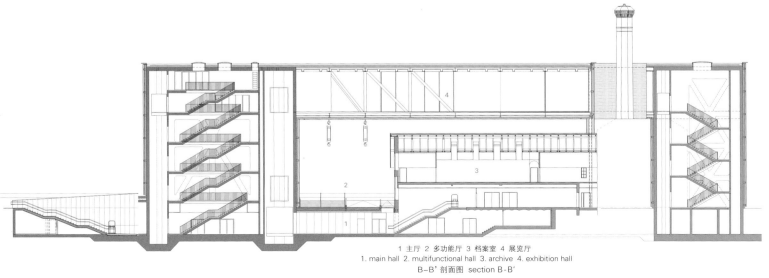

1 主厅 2 多功能厅 3 档案室 4 展览厅
1. main hall 2. multifunctional hall 3. archive 4. exhibition hall
B-B' 剖面图 section B-B'

Abandoning the idea of developing the entire parcel ground and relieving it for purposes such as performances, exhibitions, shows, happenings in the Vistula riverside "theater" extend the range of influence of the form and its content to the entire urban scale of the city.

The added building looks like a wrapped and mysterious "object" and adequate to its function structure which is hidden behind the packing.

The construction of the new part of the complex applies non-conventional solutions both in their material and static aspects (material: steel, reinforced concrete, glass) and the structure of buildings is resultant and derivative of the idea to form an external "theater" space and create mutual tension between the individual elements of the entire arrangement.

CRICOTEKA building has a form of two intersecting arches made of spatial reinforced concrete steel truss based on two shafts. The third pillar is a steel, double-rocker. The branch supports truss's given implementing raise equal to their constant deflection under

load and utility. Due to the complex geometry and other extensions, the raise values of the executive are different, which were included in the design of the cross. Relying on reinforced concrete prefabricated shafts was implemented through five pot bearings and one of the lens. Horizontal forces from the upper belt trusses are carried by special anchoring structures embedded in the shafts. The horizontal forces from the lower belts are passed in turn by the thrust on the walls of the shafts.

"CRICOTEKA" was established as a public utility institution and as such reserved the right to constitute a characteristic or even a dominant feature in the panorama of the city. The scale of the building – appropriately balanced – may reduce the negative influence of the surrounding tall buildings. The assumed height is similar to the height of the building in its background and much lower than the hotel recently built in the neighborhood.

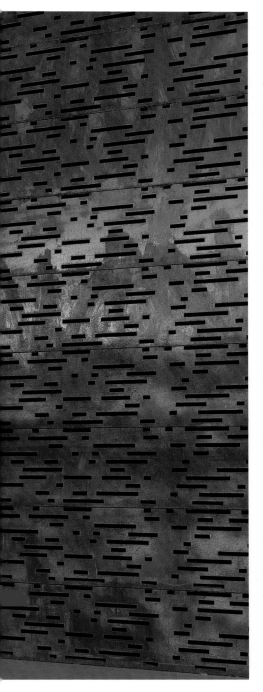

格旦斯克莎士比亚剧场

Rizzi-Pro.Tec.O

这个项目融合了两个基本概念：其一是体现历史性，其二是体现政治文化性。从历史方面来说，这座波罗的海城市早在17世纪就以其伊丽莎白女王剧院的木结构建筑而闻名。大约在四百年之后，人们又要根据几乎荡然无存的考古遗迹而在其城市和周围环境与以往截然不同的同一个地方建一座新剧院。

剧院建筑从形式上和功能上可以分为三个主要部分：外围的人行道，剧院和行政管理区。

外围的人行道是围绕着整座剧场建筑的公共通道。这个新建的行人使用的城市平台位于主入口层上方6m。这一高度为人们提供了一个全新视角来观赏既古老又现代的城市的对比和新旧结合。从功能上来说，这个外围人行道是剧场的逃生通道，连接各层，包括位于地面5m以下的地下室。

从剧场的外观来看，其轮廓主要有三个方面的特征：体量、肋状砌体以及可开合屋顶。

从体量的轮廓来看，体量分为两个截然不同的部分。首先是伊丽莎白剧院，建筑高度设定在12m。第二部分是18m高的观景塔。技术、剧场系统以及象征意义方面的需求都决定了观景塔成为最高的全景观望点。当剧院屋顶打开时，从观景塔上还能俯瞰到剧院内部。

外墙上的肋状砌体展现了剧院和观景塔的体量特征，从建筑外部显示出内部结构的模块的节奏。同时，这些外墙上的肋状砌体还能吸收屋顶开启的"两翼"对墙体施加的压力，以抵挡北风的风力。

可开合屋顶：可开合屋顶满足了建筑类型学上和象征意义上的需要。"侧翼结构"完全打开的时候高达24m，使整座建筑的垂直高度成梯度增加(6m-12m-18m-24m)。从平面上来看，建筑整体与东西向的主轴呈现出交叉的形象。观景塔把剧院和管理区横向划分开来。其中心位置重新限制了外围人行道的走向。独立于外部边缘的部分具有象征意义，用以表达不同建筑形式系统之间的空间等级。

与外墙的庄重与紧凑感形成对照的是，建筑内部使用的是轻木。伊丽莎白剧院和门厅上的悬挂结构所使用的是两种不同的轻木。伊丽莎白剧院由源自现场考古发掘所发现的2.8m×2.8m×2.8m的模块材料建造。在剧院内每一层平面内，长的两侧分别有六个模块，短的一侧有五个模块，构成一个C形。共有51个模块，可以容纳600名观众。木柱(25cm×25cm)将钢结构包含其中，其位置与画廊的模块排列保持一致。为满足剧院的不同配置需求，伊丽莎白剧院的舞台和意大利剧院的舞台都实现了完全机械化。舞台的移动装置隐藏在舞台楼板之下。舞台都是可以移动的，几乎与屋顶的侧翼结构相呼应。在门厅空间内，悬挂的方盒子与建筑外部体量的力相互抵消。人们位于两层高的大房间可以鸟瞰不同空间，这些空间包裹入口空间，且包括从位于地下室的博物馆区域到剧院的画廊区。当侧翼结构打开的时候，阳光能够直接照射到地下室。

该空间包括剧院所有的附属活动框架，即办公室、监控室、餐厅、更衣室等。屋顶平台与外围边缘位于同一层，这里可以通向向整个城市敞开的方形屋顶，即另一处让人意外的空间，可以称为第三"舞台"。

第四个"舞台"：这里望向格旦斯科的天空，望向莎士比亚的天空，望向我们每一个人内心的天空。

Gdansk Shakespearean Theater

Two fundamental assumptions converge in the project, first is the historical nature and the second is political-cultural nature. Historically: the Baltic city had already known at the beginning of the seventeenth century for the wooden building of the Elizabethan Theater. After about four centuries the new theater on the same place, but in an urban and landscape context completely different is built, restarting from their archaeological traces that are far away from presence.

The theater building is divided formally and functionally into three main parts: the walkway around the outer edges, the theater itself, and the administrative area.

The outer edges are public passageways leading around the whole complex. This new pedestrian urban platform lies six meters above the entrance level. The height offers a new viewpoint of the historic and modern city with its contrasts and compositional counterpoints. Functionally, the edges ensure escape ways from the theater, and pedestrian links with all the levels of the complex, including that of the basement at five meters below level ground. On the outside, the theater's silhouette is characterised by three

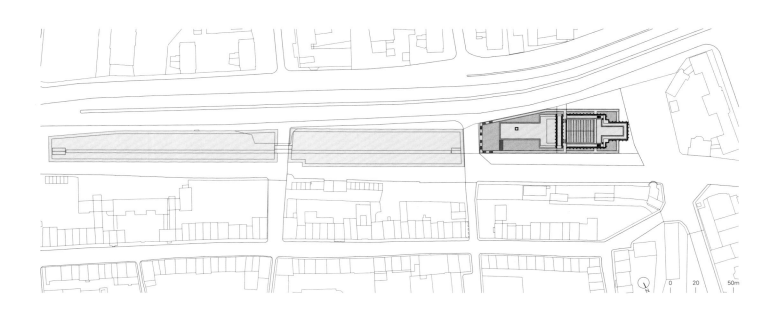

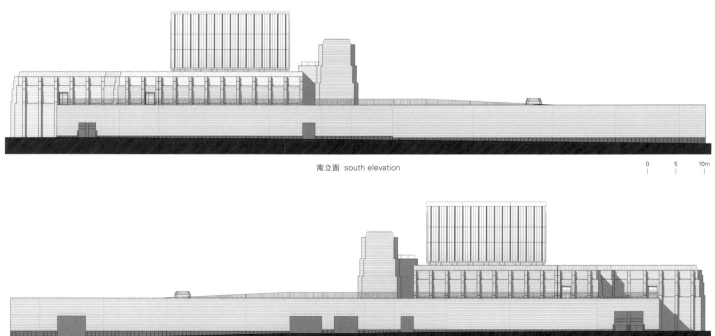

南立面 south elevation

0　5　10m

北立面 north elevation

项目名称：Gdansk Shakespearean Theater
地点：Gdansk, Poland
建筑师：Renato Rizzi
团队：Rizzi–Pro.Tec.O., Roberto Rossetto,
Roberto Giacomo Davanzo, Andrea Rossetto, Emiliano Forcelli,
Susanna Pisciella, Denis Rovetti, Lorenzi Sivieri, Luca Sirdone,
Ernst Struwig
施工单位：N.D.I
结构工程师：Armando Mammino
机械工程师：A.C.R.
顾问：Gianfranco Rorato
合作方：Q-Arch Sp. z o.o., Robert Kuzianik, Wieslaw Socha,
Anna Socha, Karol Korycki, Andrzej Dabrowski, Jan Wachacki,
Arkadiusz Kontecki, Lukasz Pilch, Bartek Zdeb, Zbigniew Kosk,
Feliks Mikulski
电气工程师：Jan Wachacki
场地监理：Bud-Invent
甲方：GTS(Gdanski Teatr Szekspirowski)
建成面积：4,000m²
材料：dark brick, basalt, wood, Istria stone, copper
造价：EUR 25,000,000
竞标时间：2005 / 设计时间：2008.12~2010.6 / 竣工时间：2014
摄影师：©Matteo Piazza (courtesy of the architect)

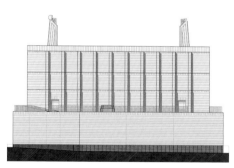

东立面 east elevation

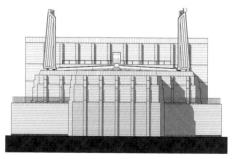

西立面 west elevation

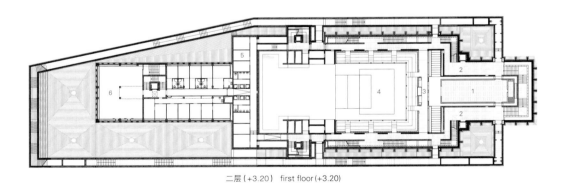

1 会议室
2 走廊
3 主任室
4 舞台
5 衣柜
6 展厅/酒吧

1. conference room
2. corridor
3. director's cabin
4. stage
5. wardrobe
6. exhibition hall/bar

二层 (+3.20) first floor (+3.20)

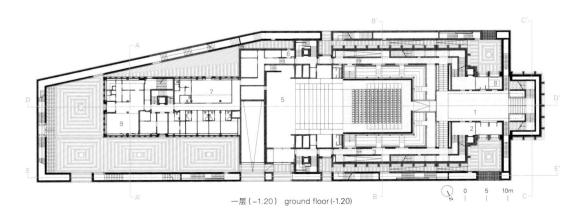

1 门厅
2 俱乐部
3 走廊
4 舞台
5 后台
6 门廊
7 温室
8 售票处

1. foyer
2. compass
3. corridor
4. stage
5. backstage
6. vestibule
7. green room
8. ticket office

一层 (-1.20) ground floor (-1.20)

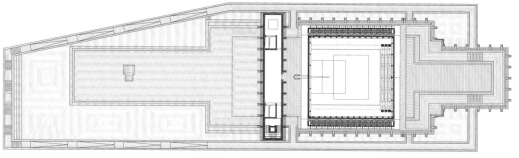

四层（+13.14） third floor (+13.14)

1 门厅
2 舞台

1. foyer
2. stage

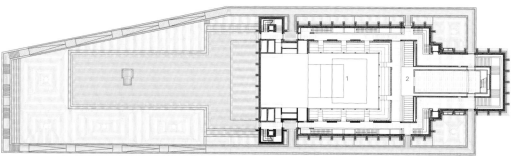

三层（+6.00） second floor (+6.00)

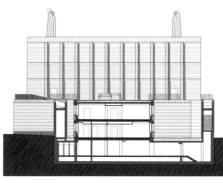
A-A' 剖面图 section A-A'

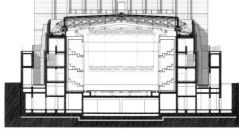
B-B' 剖面图（关闭） section B-B' (close)

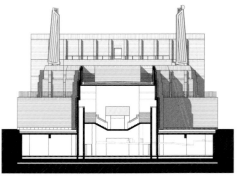
C-C' 剖面图 section C-C'

general aspects: volumes, masonry ribs, and an openable roof.
From the volume's profile two very distinct parts emerge. The first belongs to the Elizabethan theater, and sets the height of the building to 12.00m. The second belongs to the 18.00m high scenic tower. Technology, systems related and symbolic requirements make it the highest panoramic point. When the theater roof is open, the view from the tower includes the interior of the theater. Masonry ribs in the outer walls characterize the volumes of the theater and scenic tower. On the outside, they indicate the rhythm of the modular indoor structure. They did need to absorb the pressure that the open "wings" of the roof exert on the walls below in order to contrast the force of northerly winds.

Openable roof: It comes from typological and symbolic needs. With its wings opened straight up, the edges reach a height of 24 meters, concluding the vertical progression of levels (6, 12, 18, 24m). In plan, the whole assumes the figure of a cross with the main axis oriented east-west. The scenic tower transversely divides the theater area from the administrative one. Its central position restricts the pathways of the outer pedestrian edges. This figuratively autonomous part is set back from the outer edges precisely in order to express the spatial hierarchy between the different formal systems.

In contrast to the gravity and compactness of the outer walls, the interiors of the building are in light wood of two types, one for the Elizabethan theater, and one for the suspended volume above the foyer. For the former, the typological module of 2.8 x 2.8 x 2.8 meters is taken from the one found during archaeological excavations of the site. In plan, there are six modules on the two long sides and five on the one short side, making a C-shaped figure. All in all, there are 51 modules for about 600 spectators. Wooden columns(25 x 25cm) contain an internal steel structure and are positioned in accordance with the modular pattern of the galleries. The Elizabethan and Italian stages are fully mechanised to meet the theater's varying configuration needs. The stage movement facility is located in the base below the floor slab. The stages are mobile, almost in response to the wings of the roof. In the foyer volume, a suspended box offsets the external masses. A large double height room overlooks the different types of voids that envelop the entry spaces, from the museum area in the basement to the galleries of the theater itself. When the wings open, the sun rays can extend until the basement.

It contains all secondary activities of the theater: Offices, surveillance, restaurant, dressing rooms, etc. A roof terrace is at the same level with the outer edges from which it is possible to access the square roof open to the entire city, another unexpected place for representations, the third "stage".

The fourth "stage": the one that looks at the sky of Gdansk, the sky of Shakespeare, the inner sky of each of us.

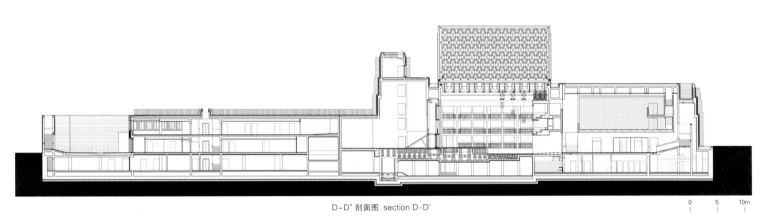

D-D' 剖面图 section D-D'

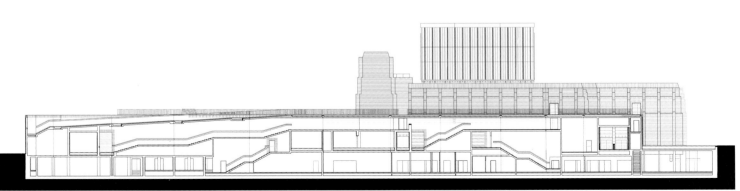

E-E' 剖面图 section E-E'

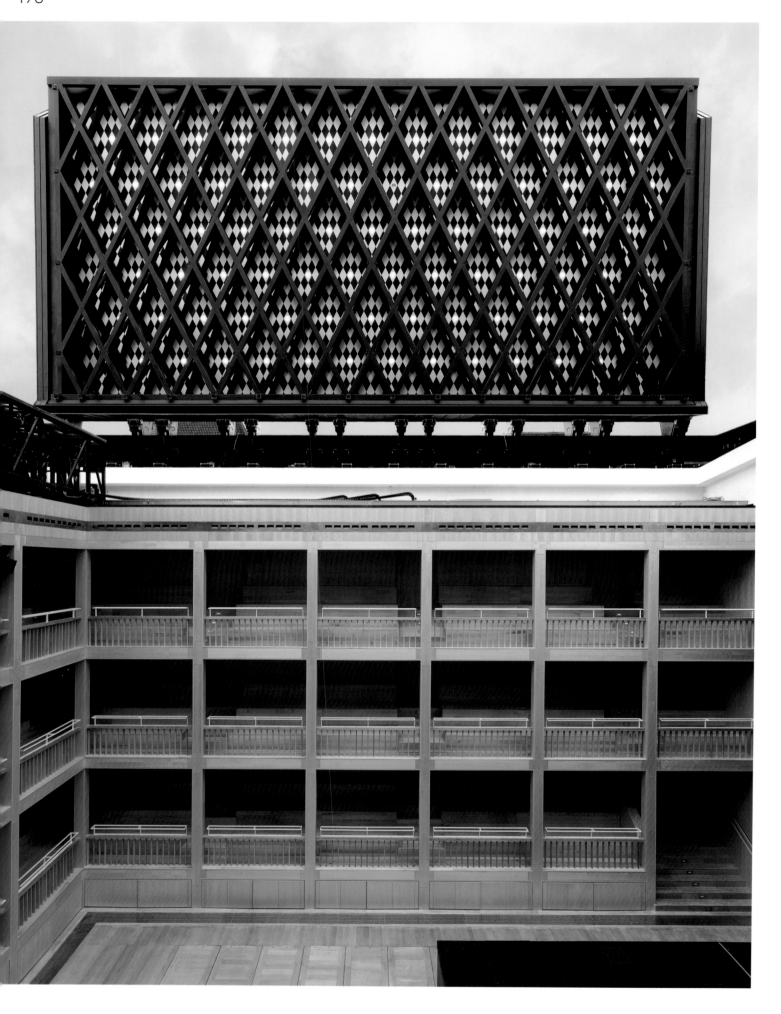

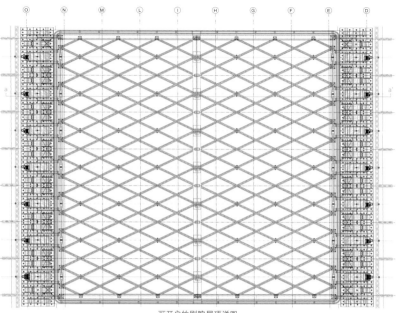

可开启的剧院屋顶详图
openable theater roof detail

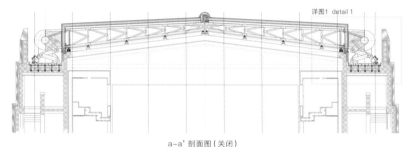

a-a' 剖面图（关闭）
section a-a'(close)

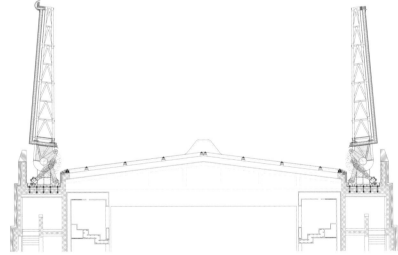

a-a' 剖面图（开启）
section a-a'(open)

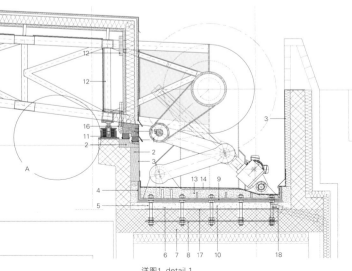

详图1 detail 1

1. double seal polymer resistant to high and low temperatures set to cellular glass
2. insulating glass foam type "foamglass T4"
3. metal cladding in green patinated copper type "tecu patina"
4. entrapment performed with epoxy type Chockfast black
5. PVC pipe protection and provisional support of the tanks
6. gusset plates
7. slab of reinforced concrete type "self compacting concrete"
8. 2270x200x20 abutment embedded in the concrete
9. ribbed steel tank 15 upwards
10. base cast lightweight concrete based on expanded perlite insulating end granulometry
11. equipment supporting elastic lattice structure
12. jet coating in lightweight concrete
13. heat- based on expanded perlite particle size of the end with formation of slope for the water evacuation, protection layer and resistant waterproofing
14. UV, the minimum thickness d 1.5 / 18mm, performed with sheath castable, polyurethane-based performed with 7 layers Sikalastic-450 and 1 layer, Sikalastic-445 profile of lamellar
15. contrast of the seals integral with the lattice structure
16. insulating panel in rigid polyurethane foam
17. centering jig
18. drainage for rain water

>>166
Rizzi - Pro.Tec.O
Renato Rizzi is an Italian Architect and Theoretician, born in 1951. Graduated from the University of Venice in 1977. Established his professional office in Rovereto(Trento) in 1978 and since 2000 in Venice. Worked for about a decade with Peter Eisenman and then came back to Italy to concern himself with teaching, design and theory. Is now an associated Professor of Architectural Theory and Design at the University IUAV, Venice.

>>22
Jean-Philippe Pargade Architecte
Jean-Philippe Pargade is an architect who graduated from the UP6 - National School of Architecture of Paris - La Villette (ENSAPLV) - and an urban planner from the National School of Bridges and Roads(ENPC). Is also a state architect board and member of the French Architecture Academy.
Created his agency in Paris in 1980 and since then, he has been involved in the building of major public projects such as research centers, teaching centers, hospitals, housing units and service buildings.

>>108
Barozzi / Veiga
Is an architectural office devoted to architecture and urbanism, founded in 2004 in Barcelona by Fabrizio Barozzi and Alberto Veiga.
Fabrizio Barozzi graduated from the University of Venice and completed his academic studies at the Sevilla School of Architecture.
Alberto Veiga graduated from the Navarra School of Architecture in 2001. Both of them have taught Architectural Design at the International University of Catalonia in Barcelona and the University IUAV of Venezia. They have won many international awards and citations for design excellence including Barbara Cappochin Prize 2011. The studio was selected to be one of the 10 firms in 2014 Design Vanguard by the Architectural Record magazine.

>>70
ZLG Design
Huat Lim graduated from the AA School under the Peter Cook, Ron Herron and Dr. Gordon Pask. Was certified for RIBA Part I and Part II and has worked for Sir Norman Foster, Zaha Hadid and Ron Herron. Has been a Managing Director for ZLG Design since its inception in 1992 and taught at the Malaya University. Susanne Zeidler hails from Frankfurt and studied art history before her postgraduate term at the Städelschule, contemporary fine arts academy under the Peter Cook, and later at the Bartlett School of Architecture, London. She has been working in Kuala Lumpur, Malaysia from 1992. Is Executive Director and Senior Partner at ZLG Design.

>>82
Boeri Studio
Was established by Stefano Boeri and Michele Brunello in Milan, 2011. Received international prizes and recognitions including International Highrise Award 2014. Studio has offices in China, Qatar, Russia and other countries abroad.
Stefano Boeri lives and works in Milan where he was born in 1956. He is committed to research and practice of contemporary architecture and urbanism nationally and internationally. He is full professor of Urban Design at the Polytechnic University of Milan and has taught courses as guest professor at various universities, including Harvard University GSD, Strelka/Moscow, Federal Institute of Technology Lausanne(EPFL) and the Berlage Institute.

>>128
AS. Architecture Studio
Was founded in Paris in 1973. Works with large international groups and develops big projects including large residential development. Has intensified its activities on the international stage. And the international presence is particularly strong in China. They have two permanent subsidiaries in Shanghai and Beijing. Define architecture as "an art committed with society, the construction of the surroundings of mankind". Its foundations lie on work group and shared knowledge, with the will to go beyond individuality for the benefit of dialogue and confrontation. Thus, the addition of individual knowledge turns into wide creative potential.

©Miguel Fernandez Galiano

>>96
Dominique Perrault Architecture
Dominique Perrault gained international recognition after winning the competition for the National French library in 1989 at his age of 36. This project marked the starting point of many other public and private commissions abroad. In 2014, he delivers the DC Tower in Vienna, the tallest tower in Austria, an icon of the new business district, as well as the Grand Theatre in Albi, France. Received many prestigious prizes and awards, including the "Grande Médaille d'or d'Architecture" from the Académie d'Architecture in 2010, the Mies van der Rohe prize, the French national Grand Prize for Architecture, the Equerre d'argent prize for the Hotel Industriel Berlier and the Seoul Metropolitan Architecture Award as well as the AFEX Award for the Ewha Womans University in Korea.

>>60
Rojkind Arquitectos
Is a Mexico City based architecture firm practicing internationally focusing on tactical and experiential innovation. Uses design thinking to cut across strategic fields looking to maximize project potential while maintaining attainability. Their multinational team works hand in hand with clients and collaboratively with experts in different fields to attain the necessary knowledge and to gain value and also to establish methods in service of the project and its areas of influence. Received worldwide recognition for its award winning projects including 2005 Architectural Record Top 10 Design Vanguard. Has been running by the founding partner Michel Rojkind and Gerardo Salinas, the partner of the firm.

>>54
Vo Trong Nghia Architects
Vo Trong Nghia graduated from Nagoya Institute of Technology with a B.Arch in 2002 and received Master of Civil Engineering from Tokyo University in 2004. In 2006, he established Vo Trong Nghia Co.Ltd.

Gabriela Etchegaray
Is a Mexican architect, born in 1984. Holds an honor degree in Architecture and Urbanism by The Ibero-American University. Completed a Master in Creative Management and Transformation of Cities in Polytechnic University of Catalonia (UPC) and UIA. Worked for three years with Architect Mauricio Rocha and co-founded an art and architecture studio, Ambrosi Etchegaray in 2011 with Jorge Ambrosi in Mexico City. Was granted from the National Fund for Culture and Arts(FONCA) as a Mexican Young Creator 2014-2015. Received Emerging Voices 2015 from the Architectural League of New York.

>>38
Luciano Pia
Was born in 1960 and graduated from Polytechnic University of Turin, School of Architecture in 1984. Has worked abroad from 1990 to 2000 and lectured ecologically sustainable buildings at Turin Polytechnic, Milan Polytechnic and at Laval University, Curtin University. He pays particularly attention to the distinctive features of the context but he also carries out new building plans to improve energetic efficiency and minimize the impact on the environment.
From 2000 he operates in his personal Architectural Studio in Turin.

>>10
Chartier Dalix Architects
Was created in 2006 by Pascale Dalix[left] and Frédéric Chartier[right]. Their projects reveal a marked sensitivity to the complexity of uses and contexts: they always propose relevant, innovative and generous solutions in terms of both formal and programmatic consistency. They have received various awards such as the Europe 40 under 40 Award(2012), and the "prix de la première œuvre".

Alejandro Hernández Galvez
Is an architect and editor based in Mexico City. Works as editorial director of Arquine and has published his articles in several magazines and journals from Mexico and abroad. Is a co-author of 100x100, 100 architects of the 20th century in Mexico and Shadows, umbrellas and hats: of architectural principles.

>>134
Renzo Piano Building Workshop
While studying at Politecnico of Milan University, Renzo Piano worked in the office of Franco Albini. After graduating in 1964, he started experimenting with light, mobile, temporary structures. Between 1965, and 1970, he went on a number of trips to discover Great Britain and the United States. In 1971, he set up the Piano & Rogers office in London together with Richard Rogers. From the early 1970s to the 1990s, he worked with the engineer Peter Rice. Renzo Piano Building Workshop was established with 150 staff in Paris, Genoa, and New York.

Angelos Psilopoulos
Studied architecture at the School of Architecture, Aristotle University of Thessaloniki(AUTh), then moved on to his Post-Graduate studies at the National Technical University in Athens(NTUA). Is currently pursuing his Ph.D. at the NTUA on the subject of Theory of Architecture, studying gesture as a mechanism of meaning in architecture. Has been working as a freelance architect since 1998, undertaking a variety of projects both on his own and in collaboration with various firms and architectural practices in Greece. Since 2003, he has been teaching Interior Architecture and Design in the Department of Interior Design, Decoration, and Industrial Design at the Technological Educational Institute of Athens(TEI).

>>118

Atelier d'architecture King Kong

Is Bordeaux-based architecture agency, established by four architects - Paul Marion[second], Jean Christophe Masnada[fourth], Frédéric Neau[third] and Laurent Portejoie[first] - in 1994 when they decided to cooperate at the end of their studies at the Bordeaux School of Architecture. They have since been entrusted with both private and public contracts in the field of architectural design and project management. They are especially strong in the field of cultural amenities in France.

>>150

Wizja sp. z o.o.

Stanislaw Denko[left] was born in 1943. His wife, Iwona Denko, is also an architect, who also acts as the office's partner. Having obtained his diploma in 1967, he worked at the Institute of Urban Studies and Spatial Planning of the Faculty of Architecture of the Krakow University of Technology, until 2000. Since 2004, he has been teaching at the Andrzej Frycz Modrzewski Academy in Krakow. Was nominated to the European Union Prize for the Contemporary Architecture – Mies van der Rohe Award 2015.

nsMoonStudio

Co-founder of nsMoonStudio, Agnieszka Szultk[right] and Piotr Nawara[middle] were born in Kraków, Poland and graduated from the Kraków Academy of Fine Arts. Agnieszka Szultk was born in 1971 and obtained a Diploma in the studio of Prof. B. Borkowskiej in 1995. She is involved in architecture, interior architecture, industrial design, exhibition. Piotr Nawara was born in 1970. He is Representative of Giugiaro Design in Poland and involved in architecture, photography, film, exhibition and industrial design.

C3, Issue 2015.7
All Rights Reserved. Authorized translation from the Korean-English language edition published by C3 Publishing Co., Seoul.

© 2016 大连理工大学出版社
著作权合同登记06-2016年第22号

版权所有·侵权必究

图书在版编目(CIP)数据

灰色建筑中的绿色自然：混合型建筑设计 / 韩国C3出版公社编；于风军等译. — 大连：大连理工大学出版社，2016.6

书名原文：C3: Green in Grey: Architecture in Hybrid Mode

ISBN 978-7-5685-0406-5

Ⅰ. ①灰… Ⅱ. ①韩… ②于… Ⅲ. ①建筑设计 Ⅳ. ①TU2

中国版本图书馆CIP数据核字(2016)第131699号

出版发行：大连理工大学出版社
　　　　　（地址：大连市软件园路80号　邮编：116023）
印　　刷：上海锦良印刷厂
幅面尺寸：225mm×300mm
印　　张：11.5
出版时间：2016年6月第1版
印刷时间：2016年6月第1次印刷
出 版 人：金英伟
统　　筹：房　磊
责任编辑：许建宁
封面设计：王志峰
责任校对：高　文
书　　号：978-7-5685-0406-5
定　　价：228.00元

发　行：0411-84708842
传　真：0411-84701466
E-mail：12282980@qq.com
URL：http://www.dutp.cn